Haystack Full
of Needles

Haystack Full
of Needles

*A memoir of research on mechanisms
of memory in the decades that
defined neuroscience*

Louis Neal Irwin

Library of Congress Control Number: 2020921821
ISBN: Hardcover   978-1-6641-4061-5
                 Softcover   978-1-6641-4060-8
                 eBook   978-1-6641-4059-2

Print information available on the last page.

Rev. date: 11/17/2020

To order additional copies of this book, contact:
Xlibris
844-714-8691
www.Xlibris.com
Orders@Xlibris.com
819888

CONTENTS

EARLY SEVENTIES (NEW YORK)

MIDSEVENTIES (DETROIT)

LATE SEVENTIES (BOSTON)

POSTLUDE

To Carol
My partner in science and life, with love

Preface

In 1960, there was no Society for Neuroscience (SfN), no Neurosciences Research Program (NRP), and neither an International (ISN) nor American (ASN) Society for Neurochemistry. By 1980, the SfN had a membership of over seven thousand, the ISN had about one thousand, and the ASN's membership approached eight hundred. Also, by 1980, the NRP had held three intensive study programs of several weeks each that had consolidated its role as a major catalytic force in the development of neuroscience.

In 1960, the structure of DNA and RNA were both known, and RNA was presumed to carry intermediate information that converted the sequence of four different nucleotides in DNA into the sequence of twenty amino acids in a protein; but how the information was transformed and how the assembly of amino acids into proteins took place was a mystery. By 1980, the genetic code for specifying all twenty amino acids had been deciphered, and the process of protein synthesis was well understood.

In 1960, the amino acid sequence of only one protein (insulin) and two peptides (oxytocin and vasopressin) was known. By 1980, thousands of proteins and peptides had been sequenced.

In 1960, the mechanisms for memory storage in the brain were totally unknown. The brain was seen as a conglomerate of hardwired circuits, though some mechanism of plasticity was presumed to exist at some level. The units of experiential information in the brain were variously believed (with little empirical evidence) to be in either a molecular or electrophysiological form, encoded at either very specific and nonredundant loci or in broadly dispersed and imprecise patterns throughout the brain, definable only in terms of statistical probabilities. By 1980, the unit of experiential information was still uncertain, but the involvement of broad areas of the brain in the storage and retrieval of memory was accepted. While the idea that chemistry alone could encode behavioral information as it did genetic information was dead by 1980, the neurochemical function and plasticity of synaptic connections between nerve cells was understood in considerable detail, providing the broad outlines and some elementary examples of how plasticity (hence learning) in the brain could occur.

In 1960, most scientists working on the brain—including those studying the neurological basis of behavior—identified as professionals in the academic fields of their training: as psychologists, anatomists, physiologists, biochemists, or pharmacologists. By 1980, researchers on the brain were more likely to be combining approaches from two or more of those classical fields and identifying as members of the newly amalgamated field of neuroscience.

In short, neuroscience coalesced as a unified field of study during the 1960s and 1970s. This occurred primarily in North America and Europe, though labs throughout the world contributed.

These same two decades saw an increase in turmoil around the world, especially in the mid-East, on the Asian subcontinent, and in Southeast Asia. The regional war in Vietnam embroiled the United States in particular as the 1960s and early '70s progressed. This coincided with a dramatic advance in the civil rights movement at home and an uptick in violence as minority aspirations, so long suppressed, erupted in frustration across the land. Domestic tranquility was further strained by growing opposition to the war in Vietnam.

I entered my senior year in high school in 1960, determined to become a scientist because I loved the enticing hints of the lifestyle of scientific research that I had tasted. During the coming decade as neuroscience was consolidating from different fields and coming into focus for me personally, I labored through years as a student, became a husband and parent, settled on and pursued scientific goals, and strove to achieve my ideal fusion of teacher and researcher. Like the world around me, those personal experiences were turbulent at times, and nothing like the lockstep successes that we read about every October when the Nobel Prize recipients are announced.

Unlike the majority of characters who appear in this book, I wasn't educated at an elite school, didn't work as a postdoc with a Nobel laureate, or spend the formative years of my academic career at cutting-edge institutions like the National Institutes of Health. Nor did I achieve the fame that most of them enjoyed. But I did get to know many of them and respect nearly all of them and build what successes I did enjoy on their accomplishments and those of others, as all scientists do. In this regard, of course, I am much more typical of the thousands of scientists who do their part in advancing knowledge without notable recognition than the much smaller number who enjoy spectacular success.

This book is written to show how individual goals intertwine with the technologies at hand to push scientific knowledge forward, often erratically, and always in the context of social forces and personal ups and downs. My sister once said to me, in effect, "I get it that you're a scientist, but what do you actually *do*? What are your days like? How do you spend your time?" Fortunately for my sister, I did not then and will not now itemize the long hours of monotony that most work in the lab or field actually entails. But I was inspired by her question to try to explain what science is about in the larger sense, how it is carried out, and why those who pursue it consider themselves so fortunate.

My story is intricately entangled with the growth of the neuroscientific and neurochemical organizations that were created between 1960 and 1980 and the scientists who developed them as they made significant advances in understanding how the brain works. By luck and circumstance, I got to know most of them and have endeavored to tell their stories in broad outline alongside my own, both before and after we became acquainted.

Because this is a true story, intended as an authentic historical document, I have tried to be as accurate as possible. Throughout the early years of my career, I kept a detailed journal of my daily activities, which I've drawn upon in writing this book. My other major source of information has been detailed interviews granted to me generously by Marc Abel, George Adelman, Bernard Agranoff, Samuel Barondes, Robin Barraco, Edward Bennett, Floyd Bloom, Rodney Bryant, Bill Byrne, Dennis and Nancy Dahl, Dominic Desiderio, Adrian Dunn, Arnold Golub, Avram Goldstein, Dianna Johnson, David Malin, James McConnell, James McGaugh, Mark Rosenzweig, Fred Samson, Richard Thompson, Alberte Ungar, and John Wilson. Autobiographical accounts by several of the major characters provided additional and confirming information. They, and other published sources, have been cited in the text.

PRELUDE

1

Could This Be the Way a Career Ends—in DuBois, Pennsylvania?

At ten o'clock in the morning on the last day of 1976, with the temperature at five degrees above zero in a stiff wind, my wife, Carol, our son, Anthony, and I drove out of our driveway in Detroit onto the Lodge Freeway and the interstates beyond that would take us away from Michigan forever. Carol and Anthony rode in the Mustang—the first car Carol and I had bought together—and I drove the VW camper accompanied by the family cat. The heater in the Mustang was functional; the one in the camper was not, and the cat demanded a transfer to the Mustang before we had reached Toledo. For ten frozen, grueling hours, we drove across northern Ohio, reaching as far as DuBois, Pennsylvania, well after dark. There in a Holiday Inn, with Carol and Anthony already asleep, I soaked in a hot bath and reflected on what had been, on the whole, a depressing year. Boston would be a bare reprieve, but the script of my career was not playing out the way I had composed it in the heady days of Houston ten years earlier or the toil and triumph of graduate school in Kansas. Whatever I had thought I would be doing by seven years after my PhD, it was certainly not going to bed without a party on New Year's Eve in a motel in DuBois, Pennsylvania.

By the time I had earned my doctorate at the age of twenty-six, I had published two papers (one in *Science*), been awarded a National Science Foundation predoctoral fellowship, worked with a world-renown neuropharmacologist, and befriended a number of the most eminent neuroscientists in the nation. Within another year, I would publish the first of numerous papers on the biochemistry of learning and memory, fully expecting to become a leader in that exciting field, and would secure a tenure track faculty position in New York City. Now on a dismal, frozen day less than seven years later, having lost my second faculty position in four years, I was driving across the Midwest with no prospect in sight of another research lab for pursuing my dream of research on the mechanisms of memory or a faculty position to fulfill my love of teaching. Where had it all gone off the rails?

* * *

2

On the Brink

As the second half of the twentieth century unfolded, breakthroughs in science and technology were quickening. The decade of the 1950s would see the launch of the space race; the development of the computer; proliferation of television; scientific study of human sexuality and research leading to birth control; the prospect of eradication of infectious diseases, including the scourge of polio; and the advent of drugs finally capable of treating the worst problems in mental health. To be sure, technology offered the prospect of a frightening future as well, with acceleration of the nuclear arms race. But scientists in general were held in high regard; and especially the launch of Sputnik by the Soviet Union in 1957 goaded the nation into a crash effort to educate more scientists, mathematicians, and engineers.

First Taste of the Rare and Sublime

Highlands High School in San Antonio, Texas, opened with a lot of fanfare and an unfinished auditorium in September 1958. Built at a cost of three million dollars on a scenic hilltop commanding the southeast quadrant of the city, it was the pride of the San Antonio public school system—one of the first of a wave of modern high schools built across the country as the postwar baby boom groped toward adolescence. At the age of fifteen, I was part of the swell but, prior to its crest, a member of that generation in transition between Eisenhower and Kennedy, slide rules and calculators, Elvis and the Beatles. I was one of the first two thousand students through the doors of Highlands High School, interested above all in football, but already aware of my limitations as an athlete; interested in journalism from the inspiration of a daily diet of Jim Bishop columns over many years; and interested especially in science because of the way the world was moving at the time.

The formative event of my teenage years, as far as my future career was concerned, was the launch of Sputnik in October of 1957—my last year in junior high. I remember hearing the first news bulletins as I pasted together a scrapbook on the football season then underway. With an artificial satellite orbiting overhead, I sensed that the world had fundamentally changed that evening, and I wanted to be a part of those who would build the brave new world that was certain to follow. Like so many of my contemporaries at that idealistic age, I was impressed with and inspired by the apparent power and promise of science and technology.

Although the space program provided an ongoing melodramatic reinforcement of the glamour of science, as I advanced through high school, the more intimate and small-scale manipulations of chemistry commanded an increasing share of my interest. At Highlands High School, with its modern chemistry lab and Homer Jackson, a truly absentminded professor who taught science as a joyful experience, I finally had a science course in school that matched my romantic image of science as it seemed to be happening in the larger outside world.

With leftover chemicals and discarded glassware from my high school lab, I began to assemble a laboratory at home, where I could fiddle into the night with projects outside the conventional chemistry curriculum. My mother watched in dismay as my bedroom evolved into a chamber for culturing fungi, heating breadcrumbs to the point of combustion, testing soil from neighborhood gardens, and like activities suggesting that her son had a curious mind in more ways than one.

By the time my chemistry course had ended, I had managed to complete a term project on the mechanism of bleaching, using my mother's various household bleaches and detergents. My conclusions were not startling, but the project was important because it confirmed for me the particular pleasure of combining mental puzzles with manual dexterity (the essence of science) in a reclusive, self-motivating, individualized activity (the nature of research).

The following year in physics, I had a more dramatic experience that left no doubt in my mind that research was what I wanted to do with my life. The project this time was to decipher the natural laws of diffusion: What determines the rate at which two solutions brought together without stirring would merge into one another? To the intellectual puzzle and the manual manipulations of this project was added another element of research I had not seen before but came to recognize frequently in subsequent years—the element of aesthetic pleasure. The solutions I worked with had to be colored differently so that I could follow the progression of one into the other. Thus, the walls of my home lab came to be lined with test tubes of multicolored solutions, forming a kaleidoscope of colors that I would occasionally lean back and look at with the eye of a self-satisfied artist.

Later I would see that while research may be relatively reclusive, it is seldom solitary. Partly this is because multiple minds are nearly always more effective that one. But another reason is the pleasure that comes from sharing the burdens of the work, the sense of fulfillment when it succeeds, and the frustrations when it doesn't. My partner in this project was Dorothy Haecker, the smartest student in the class and a good friend who lived across the street. The proximity of our houses made working together or in shifts at all hours feasible—a distinct advantage as the deadline approached and our data multiplied without giving us insight into what the factors were that governed diffusion. Finally, in the early hours of the morning just two days before our report was due, Dorothy and I saw the pattern emerge from the information we had collected. A couple of formulae were drafted that seemed to explain everything. It was a moment of genuine and sudden insight; the rarest and most sublime experience that the intellectual life can offer. It mattered little that less than half a year later, Dorothy and I would discover in our freshman college chemistry courses that the mathematical laws for diffusion we had "discovered" had been known for a century or more. That May morning in 1961, we might as well have been Albert Einstein and Madame Curie. Humility would come later, but the pleasure of the moment, intensified by fatigue, was sufficient to convince me that scientific research had to be the most rewarding of all professions. What else combined intellectual challenge, physical facility, individual initiative, aesthetic pleasure, and human companionship in a worthy venture with significance transcending the participants themselves? And what could be a more satisfying feeling than exhaustion in a worthy cause?[1]

The Known and the Unknown

As the 1960s began, I knew what DNA was, but was totally unaware of the dramatic research then unfolding that discovery of its structure had ignited. I was very interested in behavior—in part because of the mysteries of adolescent psychology that plagued me and my friends on a daily basis—but I had almost no knowledge of the brain, its composition, or ideas about how it works. The following is the essence of what I didn't know then, but would learn before long:

The brain consists of a vast network of individual cells called neurons that transmit waves of bioelectrical excitation when stimulated to do so by the release of chemicals called neurotransmitters from other neurons. Sensory neurons at the body's periphery can be activated by physical stimuli, like sound, touch, and light; and this sensory information flows upward through parts of the brain that mediate emotions, stoke or satisfy motivations, and provide a way for the environment to be reflected in consciousness. I would not have known, nor did anyone else know, how conscious awareness arises from physical and chemical processes in the brain. More fundamentally, no one was sure which processes in the brain correlate with the elements of experience— whether it was a sequence of specific neurons through a hardwired series of connections that give rise to the image of a baseball, say, or an imprecise but statistically defined and dispersed aggregation of excited or inhibited nerve cells perceived as the color blue, or a diffuse field of electromagnetic motion coursing through the fluid space surrounding neurons and their mysterious (at the time) companion cells, called glia, that evoked a feeling of depression.

I did know enough about DNA at the time to realize that all the genetic information that a single cell or a whole organism inherits from its predecessors is contained within the structure of that molecule. I had only the vaguest notions of what proteins did or how they work to build a cell, catalyze a reaction, or modify the actions of other cells, though that much of biochemistry was known by others. While everyone knew that genetic information was coded for by DNA, and as scientists were about to figure out how immunological memory is induced in a clone of cells that can last for a lifetime, no one knew whether learned behavior or the memory of lived experiences, like genetic and immunological information, is recorded in a molecular code. The attempt to answer that question would be a driving force in the transformation of neuroanatomy, neurophysiology, and biochemistry of the brain into the hybrid field of neuroscience in the next two decades to follow.

Dawning of a New Integration

How that hybrid field gave rise to what would become one of the largest and most influential scientific societies in the world—the Society for Neuroscience—had its origins at an international meeting of neurophysiologists in Moscow in 1958. Recognizing that the anatomy, physiology, and chemistry of the nervous system was a growing area in the biological sciences increasingly crossing traditional disciplinary boundaries, attendees at the Moscow meeting decided to form an international collaborative of researchers to improve communication and promote international cooperation among scientists interested in neural function. They laid the groundwork for the International Brain Research Organization (IBRO), which was formally chartered as a UNESCO organization in 1960. From IBRO would spring the seeds for the Society for Neuroscience (SfN).[2]

1   Sam Barondes expressed the same view in reminiscing about his days as an
     intern: "My 2 years at the Brigham [Hospital in Boston] . . . were filled with
     many . . . comradely experiences that come when a small group of young people
     keep working to exhaustion for a worthy cause" (Barondes, S. H. 2006. Samuel
     H. Barondes. In The History of Neuroscience in Autobiography, edited by L.
     Squire. Washington, DC: Society for Neuroscience, p. 8).

2   Neuroscience, Society for. 2019. Chapter 1: "Neuroscience before neuroscience,
     WWII to 1969." Society for Neuroscience 2019 (cited 17 Sept. 2019). Available
     from https://www.sfn.org/About/History-of-SfN/1969-1995/Chapter-1.

3

Information in the Brain

Fascination with the possible mechanisms of memory extend back at least to the start of the scientific revolution. With the relationship between brain and behavior still largely a mystery, however, and prior to an understanding of the structural organization of the brain or the physiology of its elements, mechanistic theories of how the brain stores and retrieves a record of the animal's experiences were largely absent and, when offered, were unpersuasive.

By the start of the twentieth century, that had started to change. Some point to the demonstration in 1870 by Gustav Fritz and Eduard Hitzig that nerves originating in the cerebral cortex control muscle movement of the limbs, establishing empirical verification of a mechanistic link between brain and behavior. Others justifiably cite the insights of William James in his monumental publication *Principles of Psychology* in 1890. When Sir Charles Sherrington demonstrated the integrative nature of even the simplest reflexes in the 1906 publication of *The Integrative Action of the Nervous System*, a mechanistic view of the brain's control of behavior became distinctly plausible.

Most scientists who identify themselves as neuroscientists today, though, mark the beginning of the modern era of brain science from the monumental work of Santiago Ramón y Cajal between 1890 and 1910. Best known through the French translation of his work in *Histologie du sysèm nerveux de l'homme et des Vertébrés* (1909–1911), Cajal's treatise is now available in English translation.[1] Cajal was not the first to enunciate the neuron doctrine—the assertion that the functional unit of the nervous system is the single-celled neuron, which is discontinuous from other neurons and acts through contact with them through their point of contact, the synapse—but he was the one who demonstrated it beyond question through exquisite drawings of what he observed in microscopic sections of the brain and nervous systems of many animals.

Deterministic Theories and Connectionism

The impact of Cajal's exquisite drawings of discrete neurons with their full-blown distinctive branches, spines, and nerve endings cannot be overstated. In 1871, the Italian physician and scientist Camillo Golgi published the first pictures of brain tissue stained by a new technique that visualized single nerve cells. The discreteness of the cellular elements dealt a serious blow to the then fashionable view that the brain consists of a spongelike syncytium of indistinct and continuously interconnected cytoplasmic elements. Cajal introduced technical improvements in Golgi's method and used it to explore all regions of the nervous system of many vertebrates and some invertebrates. In every case, the technique revealed discrete cells in contact with one another but enclosing delimited cytoplasmic contents, solidifying the triumph of the alternative view of nervous system organization: that its functional elements are independent, discontinuous single cells.

These developments in the visualization of brain cells coincided with the invention of the telephone, which led to the first telephone switchboards in the late 1870s. Once the neuron doctrine established single cells as the functional units of the nervous system, an analogy with

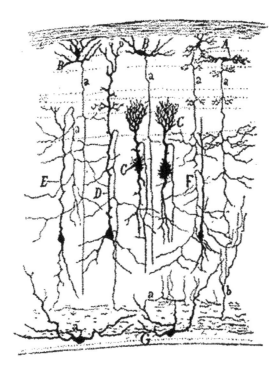

Fig. 1. Drawing of a Golgi-stained section through the optic tectum of a sparrow, from "Structure of the nervous centres of the birds" by Santiago Ramon y Cajal (1905).

the flow of information through discrete telephone lines became irresistible. If communication between two persons depended on connecting the specific telephone lines belonging to the two conversants, information flow through the nervous system could be envisioned as the sequence of specific neurons through which excitation would flow from one source of input to a specific output. And just as the telephone switchboard enabled the flexibility of connecting different parties to one another, so the switching properties at the neuron's point of contact with the next neuron in the chain could alter the destination of excitation in the nervous system. This logic provided the basis for the assumption that information in the brain is represented by activation of specific hardwired neuronal pathways, and that plasticity in that processing would involve rerouting the sequence of activation through alternative neuronal pathways. Memory, therefore, would consist of retrieving (reactivating) the specific sequence of neurons that encoded the experience in the first place.

Through much of the early twentieth century, the logic of hardwired circuits as the repository of memory, and the presumed plasticity of interneuronal (synaptic) connections as the mechanism for creating new pathways through fixed circuits that occurred during learning, dominated ideas about the storage of experience and the capacity to record new experiences. The notion that thoughts, images, and qualitative experiences depend upon specific patterns of neuronal activity was made concrete by Donald O. Hebb in 1949. He proposed a "neuropsychological" theory in which sensory stimulation activates a discrete set of neurons in a specific spatiotemporal pattern: the *cell assembly*. This is the unit of perception, as Hebb conceived it, at the multicellular level in the brain. Complex perceptions are based on the association of cell assemblies into a *phase sequence*. The association of cell assemblies is made repeatable (learning) and retrievable (memory) by changes in the synaptic efficiencies within specific neuronal circuits underlying a phase sequence. How this could be achieved was envisioned by Hebb in the following way:[2]

> When an axon of cell A is near enough to excite a cell
> B and repeatedly or persistently takes part in firing it,
> some growth process or metabolic change takes place in
> one or both cells such that A's efficiency, as one of the
> cells firing B, is increased.

The lucid images of Cajal's Golgi-impregnated neurons and the clarity of Hebb's proposed mechanism for consolidating neuronal pathways provided a compelling formulation for the representation of information in the brain as excitation through hardwired circuits, which, when learning occurred, could be modified to establish new associations. Thus, the *deterministic* representation of information in hardwired networks, modifiable through changes in the *connectivity* within the network, became the reigning paradigm of how information is represented in the brain and how learning occurs.

Statistical Theories and Dispersed Fields

While a deterministic formulation dominated the thinking of those searching for how information is expressed in the brain, there were other currents in psychology during the same period that could not be easily explained by strictly deterministic pathways. These included theories that emphasized the role of insight in problem solving, the focus on whole-pattern perception by the Gestalt psychologists, and the highly distributed nature of brain activity associated with any brain state. The latter was emphasized most notably by the classic experiments of Karl Lashley in the 1920s, in which he studied the degree to which memory in rats was degraded by surgical removal of different regions of the brain.[3] What he found in general was that the degree of memory deficit was more closely related to the amount rather than the region of the brain that was removed. While Lashley did not discount the reality of functional localization, he did point out that at the very least, mechanisms of memory must involve very widespread brain activity, with no uniquely critical localization of the memory trace.

The one technological advance that arguably was as monumental as Cajal's neuroanatomical revelations was the discovery of animal electricity. The fact that nerves and muscles could be activated by electrical currents was discovered through experiments by the Italian scientist and philosopher Luigi Galvani and his wife, Lucia, in the 1780s. When several European scientists showed in the late 1800s that weak electrical currents could be detected from the surface of the exposed cerebrum of dogs, rabbits, and monkeys, the ability to monitor brain activity through electrophysiological detection was established. The German physiologist and psychiatrist Hans Berger recorded the first human electroencephalogram (EEG) in 1924, showing organized brain waves generated spontaneously with a frequency of roughly ten cycles per second.[4] Over time, it was revealed that EEG patterns result from the integrated activity of millions of neurons driven by pacemakers in subcerebral brain centers and that the patterns vary with the functional state of the brain—being lower in amplitude and higher in frequency in alert animals and less frequent with greater amplitude during sleep, for instance—but never being absent.

While these large-scale electrophysiological events indicated a constant degree of activity over an expansive amount of cerebral tissue, they gave no evidence of the activity that generated them at the level of the individual cell. This changed with the technological development of microelectrodes that made possible the measurement of activity in single cells, beginning with the work of Edgar Adrian in 1928.[5] By the midtwentieth century, the characteristics of excitation in single cells[6] and the basic mechanisms of synaptic transmission[7] had been worked out through these single-cell unit recordings. Further advances were made by studies, such as those of David Hubel and Torsten Wiesel on the visual system of the cat,[8] showing that sensory stimulation could lead to reproducible excitation of specific neurons in the brain. Though compatible with the concept of deterministic circuitry for specific information, such studies also revealed that (1) individual neurons often are spontaneously active, and (2) stimulation through different modalities could also elicit evoked responses in the same neuron. Unlike Cajal's static images of brain cells in isolation, data from electrophysiology arose from either the integrated activity of millions of cells or the dynamic activity of individual cells that fired only in a statistically predictable way. An intermediate level of resolution was also revealed by electrodes localized to a small region of brain tissue, but not to single cells. It thus picked up averaged bioelectrical currents flowing through localized cells and their extracellular spaces—a form of activity referred to as field potentials. All three forms of neural activity—large-scale EEG waves, field potentials, and single-celled excitation or inhibition—provided a different perspective that focused on *statistical* activity across dispersed *fields* of nerve tissue.

For some neurophysiologists, it was the statistical behavior of populations of neurons that was assumed to represent information in the brain. E. Roy John was a primary advocate of this approach. For others, it was the behavior of dispersed fields of bioelectrical activity that warranted attention. W. Ross Adey was the most vigorous spokesman for this point of view.

Erwin Roy John (1924–2009) was born in Pennsylvania and grew up during the Great Depression. His studies at the City College of New York were interrupted by World War II, in which he served at the Battle of the Bulge. After the war, he completed a bachelor's degree in physics and a PhD in psychology at the University of Chicago. His research on brain function began at UCLA and continued at the University of Rochester. Then in 1974 he established the Brain Research Laboratory at the New York University School of Medicine and served as its director for thirty years. He is considered a pioneer in the field of *neurometrics*, or the science of measuring the underlying organization of the brain's electrical activity. His focus on quantitative analysis of EEG patterns across broad areas of the brain, as well as his study of multiple unit recordings in different behavioral states, provided the perspective that led him to be an early and vocal opponent of the deterministic model of information representation in the brain.

As early as 1961, John had advocated that learning needed to be studied as a process, not as an entity at a defined locus.[9] His 1967 book *Mechanisms of Memory* was the most thorough and definitive review of the subject in the second half of the twentieth century, albeit clearly biased in favor of the statistical model of information in the brain.[10] Over decades, he consistently expressed what he called a statistical configuration theory of learning, as in this paper in 1972:

> The critical event in learning is the establishment of representational systems of large numbers of neurons in different parts of the brain whose activity has been affected in a coordinated way by the spatiotemporal characteristics of the stimuli present during a learning experience.[11]

For learning to be possible, however, John admitted that some type of change had to occur in participating cells for memories to become lasting. These changes could be in synaptic efficiency, as Hebb and many—if not most—other theorists had proposed, though John suggested that other forms of plasticity were possible. In parallel with John's work, Adey was offering some ideas on those other possibilities.

William Ross Adey (1922–2004) was born in Adelaide, Australia. From the university in that city, he received degrees in medicine and surgery prior to serving in the Royal Australian Navy during World War II. After the war, he returned to the University of Adelaide for a medical degree. Fascinated with electronics all his life, he acquired an amateur radio license at the age of seventeen, designed and built the first EEG machine in Australia, and began a career of over half a century studying the bioelectrical properties of the brain and their relation to behavior. He gained international fame (and controversy) for his study of the effect of weak electromagnetic fields on biological systems, including the brain. He was also a principal investigator for NASA during the early days of spaceflight. His most salient work for the purposes of this story, however, center around his experiments and ideas about localized current flow in the extracellular spaces of brain tissue and how that current flow was affected by the interaction of ions and membrane macromolecules.

Just as John was about to publish *Mechanisms of Memory*, Adey was focusing on a tricompartmental micrometabolic module in brain tissue, consisting of neuronal, neuroglial, and extracellular compartments.[12] Decremental bioelectrical currents in the fluid surrounding those cells were influenced, in his view, by interactions between ions and macromolecules at the membrane surface. A typical observation was that injection of calcium ions into cerebral tissue caused impedance shifts in the perineuronal fluid, small and weak enough to be localized to a portion of the cell removed from the synaptic region.[13] Changes in the molecular properties of the membrane, in turn, could affect the excitability of the cell, not necessarily linked to synaptic activity.

Models Not Mutually Exclusive

In *Mechanisms of Memory*, John poses the distinction between the deterministic and statistical models as a question of whether memory is *a thing in a place* or *a process in a population*.[14] Upon reflection, it seems clear that the models address two different aspects of information in the brain. The highly dispersed nature of brain activity during the learning process, along with the probabilistic behavior of nerve cells, even when activated by the same stimulus, calls for a statistically based, dispersed field perspective. On the other hand, the need for learning to cause a definitive and essentially permanent change in the properties of neural tissue requires that some kind of physical change take place at specific sites in the brain.

From the perspective of the *deterministic-connectivity* model, Hebb was open to the view that no single synapse was necessary for the storage of the memory trace in a cell assembly. While the cell assembly theory "is evidently a form of connectionism," Hebb wrote, it doesn't "make any single nerve cell or pathway essential to any habit or perception."[15] And from the *statistical* or *field* perspective, both Adey and John recognized the need for some biophysical or biochemical alteration that had to be made at some site or sites in the brain. Adey acknowledged that the electrophysiological processes he studied most likely related to "transactional rather than to storage processes" in nervous tissue.[16] And while John's statistical configuration theory likewise emphasized how information is represented in the brain, he pointed out that the consolidation phase of memory "must be mediated by some alteration of matter, some redistribution of chemical compounds."

Assuming that some metabolic alteration must occur when a memory is created, what is the nature of the molecules involved, and what about them defines their role in memory storage? Do the memory molecules differ according to the experience that gave rise to them and therefore encode the content of the memory in their variable structures? If that is the case, how do different variants of a class of memory molecules differentially affect the function of brain cells whose collective activity represents the memory or executes its influences? Alternatively, if the metabolic consequences of memory storage derive their significance from *where* they act and *what* higher (above the molecular) level function they bring about, such as the structural alteration of a synapse or a patch of membrane surface, the structures of the contributing molecules may be invariant and therefore of no importance other than the higher level modification to which they contribute.

The chemical question boils down to this: How do molecules make a memory—by coding for experiential information directly in a purely chemical form, or by contributing building blocks to structural alterations that can only be read collectively at higher levels of organization?

1   Ramón y Cajal, S. 1995. *Histology of the Nervous System [Histologie du systèm nerveux de l'homme et des Vertébrés, 1909–1911, Madrid: Moya].* Translated by N. Swanson and L. W. Swanson. Vol. 1 & 2. New York: Oxford University Press.

2   Hebb, D. O. 1949. *The Organization of Behavior: A Neuropsychological Theory.* New York: John Wiley. The same year, Ralph Gerard proposed essentially the same idea with less detail (Gerard, R. 1949. *Am J Psychiat* 106: 161–173).

3   Lashley, K. S. 1929. *Brain Mechanisms and Intelligence.* Chicago: University of Chicago Press.

4   Berger, H. 1929. Über das Elektrenkephalogram des Menschen. I. *Arch Psychiat* 87: 527–570.

5   Adrian, E. 1928. *The Basis of Sensation.* London: Christophers.

6   Hodgkin, A. L. and A. F. Huxley. 1952. A quantitative description of membrane current and its application to conduction and excitation in nerve. *J Physiol* 117: 500–544.

7   Eccles, J. C. 1957. *The Physiology of Nerve Cells.* Baltimore, MD: Johns Hopkins Press.

8   Hubel, D. H. and T. N. Wiesel. 1959. Receptive fields of single neurones in the cat's striate cortex. *J Physiol* 148: 574–591.

9   John, E. R. 1961. High nervous functions: brain functions and learning. *Ann Rev Physiol* 23: 451–481.

10   John, E. R. 1967. *Mechanisms of memory.* New York: Academic Press.

11   John, E. R. 1972. Switchboard versus statistical theories of learning and memory. *Science* 177: 850–851.

12   Adey, W. R. 1967. Intrinsic organization of cerebral tissue in alerting, orienting, and discriminative responses. In *The Neurosciences: A Study Program*, edited by G. C. Quarton, T. Melnechuk, and F. O. Schmitt. New York: The Rockefeller University Press.

13   Wang, H. H., T. J. Tarby, R. T. Kado, and W. R. Adey. 1966. Periventricular cerebral impedance after intraventricular injection of calcium. *Science* 154:

1183–5.

[14] John, E. R. 1967. *Mechanisms of Memory.* New York: Academic Press; p. 17.

[15] Hebb, D. O. 1949. *The Organization of Behavior: A Neuropsychological Theory.* New York: John Wiley; p. xix.

[16] Adey, W. R. 1969. Slow electrical phenomena in the central nervous system: chairman's introduction. *Neurosci Res Prog Bull* 7: 79–83.

4

Molecular Biology

In the hindsight of time, the notion that all the information required for the intricate elements of mental life—from complex imagery to the details of emotions and motivation, to the memories of a lifetime—could be stored in and retrieved from a purely chemical form in the brain seems naive. But in 1950, the idea that all the complexity of an organism's heredity could be coded for in a straightforward and relatively simple molecular structure seemed just as hard to believe. The twin biological mysteries of inherited (genetic) and acquired (learned) information were equally inexplicable and not perceived as being that different. A process even closer to learned information was the phenomenon of immunity, whereby exposure to disease-causing pathogens, or almost any type of chemical substance (antigen) foreign to the body of an organism, would cause that organism to produce a protein (antibody) that neutralizes the antigen. The fact that immunity could be maintained in many cases for the life of the host indicated a permanent form of storage so like experiential memory that it came to be known as immunological memory. As information gathered through the first half of the twentieth century that both genetic information and immunological memory were chemically based, the temptation to suspect the same would be true for learned information was strong and well within the realm of conventional speculation. Those who came to criticize the earnest search for the molecules of memory during the 1960s and 1970s[1] were apparently free of the prevailing mindset at the midpoint of the twentieth century.

Genetic Information

By 1950, it was recognized that deoxyribonucleic acid (DNA) is the repository of genetic information in the cell, though how a molecule of such apparent simplicity (a long strand of alternating phosphate and sugar molecules, with only four different kinds of nucleotide bases attached to the sugar units) could hold so much information was a mystery.[2] In 1953, that mystery was largely dispelled when the detailed structure of DNA was revealed by James D. Watson, Francis Crick, Rosalind Franklin, Maurice Wilkins, and their colleagues.[3] Not only did the potential for the arrangement of the nucleotide bases to contain a huge amount of information become obvious, but the double-stranded structure of the molecule suggested a means for that information to be replicated with each cell cycle and thus be transmitted to descendant individuals.[4]

Ribonucleic acid (RNA), also containing a phosphate-sugar backbone and four nucleotide bases, is closely related in structure to DNA, but its function remained unknown through the 1950s. In 1958, Crick summarized the growing view that (1) the sequence of amino acids in proteins is coded for by the sequence of nucleotides in DNA (the sequence hypothesis) and (2) that information flows from DNA through RNA to protein, but not in the reverse direction (the central dogma, so-called by Crick because there was as yet no evidence for it).[5] Francois Jacob and Jacquez Monod proposed that structural genes (partial segments of DNA) are expressed through complementary segments of messenger molecules whose readout depended on the needs of or conditions within the cell.[6] That same year, a particular form of RNA, eventually named messenger RNA (mRNA), was shown to be an intermediate between DNA and protein, as Crick and his colleagues had predicted.[7] However, mRNA was single-stranded only and more easily modifiable than DNA, so it became the more likely candidate for storage of acquired information prior to its translation into proteins that could alter the characteristics of a nerve cell—either its excitability or connectivity or both.

Proteins could alter the properties of nerve cells in several ways. As enzymes, they could catalyze the formation of new synaptic transmitters, more of an existing transmitter, or increase transmitter breakdown. As structural components, they could provide building blocks for growing new synaptic connections, thus altering the connectivity of nerve cells. As membrane channel molecules, they could increase or decrease the flux of ions through the membrane, hence altering the excitability of the cell. The ability of a protein to do any or all those things depended on its three-dimensional structure (conformation), which was determined by the arrangement of amino acids that made up its primary structure.

Amino acids come in about twenty different varieties. When it was shown in 1961 that mRNA codes for the arrangement of amino acids in a protein, the mystery obviously became one of figuring out how the mRNA molecule with four types of bases codes for the arrangement of twenty different types of amino acids in a polypeptide.[8] First, Marshall Nirenberg and Heinrich Matthaei,[9] then Severo Ochoa[10] and his colleagues were able to demonstrate that the sequence of four different bases in mRNA read three at a time could spell out sixty-four different combinations (4 × 4 × 4 = 64), or more than enough to code for twenty separate amino acids, with some redundancy left over.

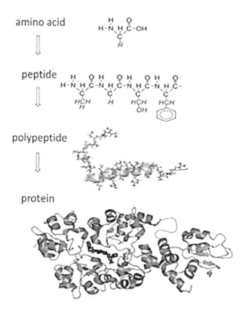

Fig. 2. Amino acids strung together become peptides. Longer peptide strands are called polypeptides, and very long polypeptides are proteins. This figure illustrates schematically how polypeptides and proteins assume three-dimensional shapes specific for their function. For instance, enzymes act to break apart molecules (like the forked serpentine structure in the middle of the protein) that fit into pockets within the enzyme molecule. (Modified with permission from Fig. 2.6 in *Cosmic Biology* by Louis Irwin and Dirk Schulze-Makuch [2011].)

Immunological Memory

Oswald Avery and Michael Heidelberger discovered in the 1920s that antibodies are proteins.[11] By 1950, building on theories dating from the turn of the century, Linus Pauling showed that antibodies bind to antigens with steric specificity, meaning that an antigen's structure is complimentary to the antibody's structure and therefore fits into it like a key into a lock. The question then became one of how a large antibody protein could turn out to be such a precise fit for the near-infinite number of antigens that nature presents to any given organism. The instructional theory held that when antigens encounter a naive antibody, the antibody molds itself somehow around the antigen and instructs the cell to produce more form-fitting molecules for future encounters with the same antigen. The selectional theory, in contrast, proposed that antibodies come from a population of cells already present that produce a large variety of antibody structures. When a cell possessing an antibody that happens to have a structure complimentary to the antigen, that cell is selected to produce more of the same antibodies for future encounters with the antigen.

Both theories faced severe challenges. The instructional theory had no explanation for how a protein molecule with a single invariant amino acid composition could be induced to change its shape, maintain the change permanently, and promote the production of more antibodies just like it. The selectional theory implied that a huge repertoire of different preexisting antibody shapes had to be producible (and therefore coded for) by genetic information already present. That an organism could carry enough information to code for a nearly infinite number of amino acid sequences capable of binding to antigens of virtually any shape seemed highly improbable. Nonetheless, through the 1950s, evidence gradually accumulated in favor of the selectional theory, implausible though it seemed.

A major advance in understanding the structure, and ultimately the mechanism of antibody binding and selection, was made in the early 1960s by Gerald Edelman,[12] Rodney Porter,[13] and their colleagues. They deduced the complete amino acid sequence of immunoglobulin G (IgG), the most prevalent antibody class produced in mammals, demonstrating that the molecule consists of two heavy chains and two light chains, which at one end of each consists of a highly variable sequence of amino acids. The other end of both chains consists of a constant (invariable) sequence of amino acids. The variable portion of the molecule allows for different antigenic shapes to be bound by a uniquely fitting three-dimensional shape brought about by the variable sequence of amino acids at the end of the antibody molecule, while the constant end enables generic functions for carrying out the antibody's method of neutralizing the foreign antigen, by agglutination (clumping together), precipitation, or destruction.

The problem of how a virtual universe of antigenic structures could be anticipated by the genome of an infected or invaded organism remained. In 1967, Edelman and Joseph Gally proposed that duplication of a relatively small number of genes could generate a pool of variable genes that, through reshuffling, could give rise to a great variety of sequences.[14] In a landmark series of experiments in the 1970s, Susumu Tonegawa confirmed this as a major mechanism for inducing variability into the antigen-binding portion of the antibody molecule,[15] and other studies in the following decades provided conclusive evidence for the selectional theory of antibody production.

Crick, Barondes, and Edelman

The purpose of this very brief overview of advances in molecular biology around the middle of the twentieth century is mainly to set the stage for the way the search for molecular mechanisms of memory unfolded during that era of great breakthroughs in molecular biology. It also serves to introduce three figures who would each go on to make an important impact in neuroscience and cross paths tangentially with my own personal story.

Francis Crick was born in 1916 and raised near the town of Southampton, England. His father was not a scientist, but his grandfather was a naturalist who had corresponded with Charles Darwin. He earned a bachelor of science degree in physics from University College, London, and began doctoral studies there that were interrupted by the onset of the Second World War, during which he worked for the Admiralty Research Lab designing mines that were effective in countering German minesweepers. At the end of the war, he thought hard about the direction of his career, narrowing his choice to either "the boundary between the living and nonliving, or the workings of the brain." By training and experience, he decided he was better suited for the former—what later would be called molecular biology.

Crick's first job in his newly chosen career path was a study of the properties of cytoplasm in chick fibroblasts (embryonic muscle cells). He was not deeply interested in this project, but it generated his first two publications and left him time to do a lot of reading. After two years, he seized upon an opportunity to join the lab of Max Perutz and John Kendrew at the Cavendish Lab in Cambridge, where x-ray studies of protein structure were more to his liking.

In 1951, James Watson joined the Cavendish Lab in quest of a means to study the structure of DNA. With similar interests and temperament, Crick and Watson became immediate friends and collaborators. That year, Linus Pauling had deciphered the structure of the alpha helix of proteins by combining x-ray diffraction data with construction of molecular models. Crick and Watson decided to use the same strategy, using data generated at Kings College, London, by Rosalind Franklin working in the lab headed by Maurice Wilkins. Their efforts led to the discovery of the double-helical structure of DNA. In 1962, Crick, Watson, and Wilkins were awarded the Nobel Prize in Physiology and Medicine for their discovery of the structure of DNA. Franklin, whose data had made the discovery possible, had died at the age of thirty-seven and thus was ineligible for her deserved share of the prize. At the same ceremony, Perutz and Wilkins received the prize for their analysis of the structure of hemoglobin and myoglobin.

It was agreed by almost everyone that the sequence of nucleotides inside the double helix must specify the sequence of amino acids in the protein which that stretch of DNA directs. Crick spent the decade following discovery of the double helix unravelling the details of how a sequence of four different nucleotide bases in DNA could control the alignment of twenty different types of amino acids in a protein. In 1955, he introduced the notion that a unique adapter molecule (later named transfer RNA, or tRNA) for each amino acid must be required for lining up amino acids in the correct order. In 1961, he proved with Sydney Brenner and their colleagues that the code for each amino acid occurred in a non-overlapping sequence of three nucleotides.[16]

In a talk given in May of 1961 at the National Institutes of Health (NIH) in Bethesda, Maryland, Brenner declared that a way to prove that RNA carries the information from DNA that orders the arrangement of amino acids in a protein would be to place an extraneous strand of RNA in a cell-free protein synthesis system (using the disrupted contents of a cell with its own native nucleic acids removed) and see if a protein could be synthesized by the system in a test tube.[17] Attending the lecture was Gordon Tomkins, recently appointed head of the new Laboratory of Molecular Biology at NIH, who had just recruited a young biochemist at the end of his three years of postdoctoral work at NIH, Marshall Nirenberg, and a postdoc to work with him, Johann Heinrich Matthaei, from Germany. That very month, Nirenberg and Matthaei set in motion the experiment that showed that an artificial strand of RNA containing only the nucleotide uridine (poly-U) led to synthesis of an artificial polypeptide consisting only of the amino acid phenylalanine.[18] The first word of the genetic code in RNA was thus discovered to be UUU, which translated to phenylalanine in the polypeptide.

Just as this experiment was in its preliminary stages, Tomkins became friends with Samuel Barondes, a recent medical school graduate serving his military obligation as a physician in the United States Public Health Service by working in the clinical endocrinology branch at NIH. One day, in a discussion of their mutual interest in endocrinology, Tomkins told Barondes that, in reality, endocrinology boils down to molecular biology. "What exactly is molecular biology?" Barondes asked. Tomkins then proceeded to describe the exciting findings in the field at the time—how DNA contains the organism's genetic information, how different segments of DNA are apparently transcribed into an intermediate messenger that orders the sequence of amino acids in a protein, how some other mechanism controls which proteins are synthesized in any given cell, and how hormones must act by regulating the production of specific proteins in their target tissues. Barondes was overwhelmed at the vision of this brave new world and decided to ask Tomkins on the spot if he could join the Tomkins lab. But Tomkins was on the brink of leaving for a sabbatical in Paris, so he suggested instead that Barondes join the new Laboratory of Molecular Biology just around the corner, where Nirenberg and Matthaei were on the verge of breaking the genetic code.[19]

Sam Barondes was born in 1933 and raised in the Brighton Beach neighborhood of Brooklyn. His education through high school was managed by rabbis in Jewish parochial schools for boys. His mother hoped he would attend Yeshiva University and become a rabbi, but he aspired to a broader world and chose Columbia University instead. There he became fascinated with the scientific study of behavior and determined to become a research psychologist. His uncle and titular head of the extended family approved of his career choice, but told him, "First, you have to go to medical school. That will broaden your horizon and provide you with some security . . . When you've finished medical school, you'll be in a great position to start doing exactly what you want."[18] So that is what he did, enrolling in the Columbia College of Physicians and Surgeons in 1954.

His years in medical school were exciting in all areas, except for psychiatry, which was devoted in the fashion of the times almost exclusively to psychotherapy—a technique based on strong opinions but no experimentation. Though intensely interested in psychopathology, his interests then veered more toward endocrinology, which was an evidence-based branch of medicine with clear clinical relevance. This steered him upon completion of his medical degree toward an internship at the Peter Bent Brigham hospital in Boston, where George Thorn had established a highly regarded research program in endocrinology. In order to complete a residency at the Brigham, he had to spend two years in a research lab, which led him to apply for and gain a coveted appointment in the clinical endocrinology branch at NIH. He arrived there on July 1, 1960, and began successful experiments on the neurochemical control of the pituitary gland. His epochal meeting with Tomkins happened in early 1961, leading to a place in the lab with Nirenberg and Matthaei one month before they discovered the first word of the genetic code.

By all accounts, **Gerald Edelman** was one of the most brilliant *and* contentious scientists of the twentieth and early twenty-first century.[20] His brilliance came to light early when, as a graduate student, he began to unravel the structure of antibodies, leading to a Nobel Prize in 1972 with Rodney Porter for their combined but independent work on deciphering the structure of the IgG molecule.

Gerald Maurice Edelman was born in 1929 and raised in the Ozone Park neighborhood of Queens, New York. His early affinity for music led him through years of violin study, but he abandoned that in favor of science upon deciding that he didn't have the drive or talent to succeed as a concert violinist. He was graduated *magna cum laude* from Ursinus College in 1950 and completed an MD degree at the University of Pennsylvania School of Medicine in 1954. Following an internship at the Massachusetts General Hospital, he joined the Army Medical Corps and served briefly in Paris, where he became interested in immunology. In 1957, he enrolled as a graduate student at the Rockefeller Institute (now University) and completed work for the PhD in 1960. He stayed at the Rockefeller, first as assistant dean of graduate studies and advancing to professor until he moved to the Scripps Research Institute in La Jolla, California, in 1992.[21]

With his colleague, Joseph Gally, Edelman discovered that the light chains of antibodies were secreted in homogeneous form in the urine of a patient with myeloma, making them amenable to structural analysis.[22] The effort to do so proceeded through the 1960s, culminating in the publication of the complete molecular structure of an IgG in 1969.[23] In the process of this work, Edelman put forth ideas about the origin and mechanism of antibody diversity in vertebrates, leading ultimately to the triumph of the clonal selection theory of antibody specificity.[24] The "memory" of an antigen is preserved by selection of a cell in the infected organism, which carries an antibody on its surface to that antigen and which rapidly reproduces a population of identical cells, or clones, carrying a large number of the same antibodies for mounting a more vigorous defense the next time the antigen is encountered.

Following his receipt of the Nobel Prize in 1972, Edelman's lab turned to the problems of cell-cell interactions, embryonic morphogenesis, and molecular recognition molecules on cell membranes. After years of laborious effort, mainly by Urs Rutishauser, Edelman's lab identified a glycoprotein on the cell surface of neurons that mediates adhesion during neurogenesis.[25]

While the antibody work was finishing and the work on cell adhesion was beginning, Edelman started turning his attention toward the brain.

By the late 1960s, he had already attracted the attention of Francis
O. Schmitt, founder of the Neurosciences Research Program, who
recruited Edelman to join the elite group of scientists that Schmitt had
brought together to study "the physics of the mind." Impressed with the
role of selectional mechanisms in solidifying immunological memory,
Edelman developed a theory of memory and higher brain function
based on specification of groups of neuronal pathways (composed of
different cerebrocortical columns of neurons envisioned by Vernon
Mountcastle to encode specific elements of sensory input) through
selectional mechanisms.[26]

1   David Hubel, whose research on visual information processing contributed
     greatly to the connectionistic view of brain function, provided a particularly
     caustic example by writing, "A few years ago the notion was advanced that
     memories might be recorded in the structure of large molecules . . . Few people
     familiar with the highly patterned specificity of connections in the brain took
     the idea seriously . . . The fad has died out, but the fact is that neurobiology
     has not always advanced or even stood still; sometimes there is momentary
     backsliding" (Hubel, D. 1979. The brain. *Sci. Amer.* 241: 45–53).
2   Judson, H. F. 1979. *The Eighth Day of Creation.* New York: Simon and Schuster.
3   Watson, J. D. and F. H. Crick. 1953. Molecular structure of nucleic acids; a
     structure for deoxyribose nucleic acid. *Nature* 171: 737–8; Wilkins, M. H., A.
     R. Stokes, and H. R. Wilson. 1953. Molecular structure of deoxypentose nucleic
     acids. *Nature* 171: 738–40; Franklin, R. E. and R. G. Gosling. 1953. Evidence
     for 2-chain helix in crystalline structure of sodium deoxyribonucleate. *Nature*
     172: 156–7.
4   Watson, J. D. and F. H. Crick. 1953. Genetical implications of the structure of
     deoxyribonucleic acid. *Nature* 171: 964–7.
5   Crick, F. H. 1958. On protein synthesis. *Symp Soc Exp Biol* 12: 138–163.
6   Jacob, F. and J. Monod. 1961. Genetic regulatory mechanisms in the synthesis
     of proteins. *J Molec Biol* 3: 318–356.
7   Brenner, S., F. Jacob, and M. Meselson. 1961. An unstable intermediate carrying
     information from genes to ribosomes for protein synthesis. *Nature* 190: 576–
     581; Gros, F., H. Hiatt, W. Gilbert, C. G. Kurland, R. W. Risebrough, and
     J. D. Watson. 1961. Unstable ribonucleic acid revealed by pulse labelling of
     Escherichia coli. Nature 190: 581–5.
8   A peptide, polypeptide, and protein all consist of a specific sequence of amino
     acids, lined up as instructed by the sequence of nucleotide bases in RNA. They

differ only in size, with peptides consisting of as few as two, and a protein with tens to hundreds, of amino acids. The term *polypeptide* is generally used when the number of amino acids in the peptide or protein is not known.

[9]   Nirenberg, M. W. and J. H. Matthaei. 1961. The dependence of cell-free protein synthesis in *E. coli* upon naturally occurring or synthetic polyribonucleotides. *Proc Natl Acad Sci USA* 47: 1588–602.

[10]   Ochoa, S. 1963. Synthetic polynucleotides and the genetic code. *Fed Proc* 22: 62–74.

[11]   Van Epps, H. L. 2006. Michael Heidelberger and the demystification of antibodies. *J Exp Med* 203: 5.

[12]   Edelman, G. M. and J. A. Gally. 1962. The nature of Bence-Jones proteins. Chemical similarities to polypetide chains of myeloma globulins and normal gamma-globulins. *J Exp Med* 116: 207–27; Fougereau, M. and G. M. Edelman. 1965. Corroboration of Recent Models of the Gamma-G Immunoglobulin Molecule. *J Exp Med* 121: 373–93.

[13]   Cohen, S. and R. R. Porter. 1964. Heterogeneity of the peptide chains of gamma-globulin. *Biochem J* 90: 278–84; Press, E. M., P. J. Piggot, and R. R. Porter. 1966. The N- and c-terminal amino acid sequences of the heavy chain from a pathological human immunoglobulin IgG. *Biochem J* 99: 356–66.

[14]   Edelman, G. M. and J. A. Gally. 1967. Somatic recombination of duplicated genes: a hypothesis on the origin of antibody diversity. *Proc Natl Acad Sci USA* 57: 353–8; Edelman, G. M. 1970. The structure and genetics of antibodies. In *The Neurosciences: Second Study Program*, edited by F. O. Schmitt. New York: The Rockefeller University Press.

[15]   Hozumi, N. and S. Tonegawa. 1976. Evidence for somatic rearrangement of immunoglobulin genes coding for variable and constant regions. *Proc Natl Acad Sci USA* 73: 3628–32.

[16]   Crick, F. H., L. Barnett, S. Brenner, and R. J. Watts-Tobin. 1961. General nature of the genetic code for proteins. *Nature* 192: 1227–32.

[17]   Judson, H.F. 1979. *The Eighth Day of Creation*. New York: Simon and Schuster.

[18]   A symposium was convened on June 4, 1961, at Cold Spring Harbor, featuring all the leading molecular biology labs actively working on the coding problem at that time. None of the then-famous workers in the field had ever heard of Nirenberg or Matthaei. Nirenberg had applied to attend but, unknown and unpublished, had been rejected (Judson, 1979, op cit.).

[19]   Barondes, S. H. 2006. Samuel H. Barondes. In the History of Neuroscience in Autobiography, edited by L. Squire. Washington, DC: Society for Neuroscience.

[20]   His son, David Edelman, once admitted in a private conversation with me that his father was regarded by some as the "dark knight of science." But that view, he implied correctly, would be a vast oversimplification of Edelman as a scientist and a man.

[21] Rutishauser, U. 2014. Gerald Edelman (1929–2014). *Nature* 510: 474.

[22] Edelman, G. M. and J. A. Gally. 1962. The nature of Bence-Jones proteins. Chemical similarities to polypetide chains of myeloma globulins and normal gamma-globulins. *J Exp Med* 116: 207–27.

[23] Edelman, G. M., B. A. Cunningham, W. E. Gall, P. D. Gottlieb, U. Rutishauser, and M. J. Waxdal. 1969. The covalent structure of an entire gamma G immunoglobulin molecule. *Proc Natl Acad Sci USA* 63: 78–85.

[24] Edelman, G. M. 1970. The structure and genetics of antibodies. In *The Neurosciences: Second Study Program*, edited by F. O. Schmitt. New York: The Rockefeller University Press.

[25] Rutishauser, U., J. P. Thiery, R. Brackenbury, and G. M. Edelman. 1978. Adhesion among neural cells of the chick embryo. III. Relationship of the surface molecule CAM to cell adhesion and the development of histotypic patterns. *J Cell Biol* 79: 371–81.

[26] Edelman, G. M. and V.B. Mountcastle. 1978. *The Mindful Brain: Cortical Organization and the Group-Selective Theory of Higher Brain Function.* Cambridge, MA: MIT Press.

EARLY SIXTIES (LUBBOCK)

5

Encoding Memory

Growing up in San Antonio, I had always imagined I would go to the University of Texas at Austin, which had grandly impressed me from childhood. The graceful Spanish architecture, the bustle of the campus, the expansive mall, and especially the soaring central library tower with its inspirational engraving across the main entrance: "Ye shall know the truth and the truth shall make you free"—these were the images that formed my early vision of the college experience. But as the actual time approached, my enthusiasm for the university in Austin waned. The fact that so many of my contemporaries were headed for Austin curiously influenced me to think of going in a different direction. I looked to northwest Texas instead, where Texas Tech was the dominant university—the third largest in the state at that time.

While I really liked the city of Austin and the beautiful hill country surrounding it, I had family roots in West Texas and found that land to have a peculiar, different attraction as well. I was born in 1943 in Big Spring at the eastern edge of the oil-rich Permian basin. Ranching, oil, and dry-land farming are the primary pursuits around this little city, which grew up during the Second World War as a refueling stop for American Airlines on its Ft. Worth to El Paso route. About an hour's drive north of Big Spring lies the southern boundary of the caprock—a geological formation that lifts northwest Texas onto a high flat plain.

Thirty miles beyond the beginning of the caprock, the four or five sizeable buildings of downtown Lubbock and the buildings of the Texas Tech campus west of downtown rise up like pegs in a pancake from the surrounding terrain.

Off to College

So I entered Texas Tech in the fall of 1961 as a chemistry major. Despite the spectacular success of the physical chemistry project that Dorothy and I had carried out as high school seniors, by the time I enrolled in college, it was the chemistry of life that appealed to me most. Unlike my high school experience with chemistry and physics, biology had been dry and boring—a purely descriptive exercise in memorization of trivia, with little attention to theory or mechanism and certainly none at the molecular level. By that time, though, the breakthrough discoveries in molecular genetics were beginning to make their way into popular science writing, which I consumed avidly. From publications such as *Science Digest*, I had learned that genes consist of long strands of DNA, which (in a way not yet clear to me) direct the manufacture of other large molecules that carry out all the functions of the cell. The modem era of psychopharmacology was dawning, and I became vaguely aware of the power of drugs to treat behavioral disorders during a summer of work at the tuberculosis sanitarium and state hospital near where I lived.

Probably my greatest motivation toward biology came from the experience of growing up with the last generation to be exposed in a major way to polio—a terrible neurological disease that cripples children so cruelly. Jonas Salk's development of an effective vaccine against polio, made possible by the pioneering discoveries of John Enders and other microbiologists, had a particularly strong and positive impact on the public's esteem for biomedical research, and on me personally. I admired Alan Shepard and John Glenn and followed their adventures into space with enthusiasm, but Salk and Enders were my role models, probing at the limits of the known universe inside the living cell.

The afternoon of September 19, 1961, I sat down in the front row of my first college class in biology. The professor walked in, tall and thin in a starched white shirt and tie.

"My name in Dr. Tinkle—like a bell."

With that, he launched the most stimulating science course I would ever take. In Don Tinkle's classroom, every lecture was interesting, every topic stimulating. As the semester progressed, my focus shifted with the subject of the week, each seemingly more enticing than the one before it.

The girl who sat beside me in biology was Penny May, a scholarship student from Waltrip High School in Houston who had come to Texas Tech with interests in English, architecture, and premedical studies. She had wanted to go to Sophie Newcomb in New Orleans, but her father had made it clear that he would not pay for his two daughters to go to college, so she settled for Texas Tech, where she managed the expenses herself through scholarships and by moving in with her aunt who lived in Lubbock. Since our questions and comments in zoology soon set us apart as the two students most involved and interested in the subject matter, we began to notice one another, to work together, and to become good friends.

It turned out, curiously enough, to be a subject that Don Tinkle dwelt little on that attracted me most. The topic, animal behavior, was worth about a single lecture near the end of the semester. Behavior is the characteristic of animals that gives them their individuality and directly defines an animal's role in the scheme of nature. Since behavior is the province of the brain, and the brain is a material substance amenable to physical and chemical analysis, I saw no reason to doubt that the greatest mysteries of life could be revealed through the logical magic and manipulations of chemistry—a labor that had given me pleasure in the past and would continue to do so, I was sure. In this way, I came to decide that my main goal in life would be the scientific study of behavior and the brain. I well remember when these thoughts first came into focus. Riding the bus home from Lubbock to San Antonio that first Christmas of my freshman year, I watched the sun set north of Big

Spring, marveled at the bleak foreboding beauty of the Texas plains, and reflected on the adventures to come.

Holger Hydén

One morning during the fall of my sophomore year at Texas Tech, I heard a brief news item on the radio that startled me to attention. According to the report, a Swedish scientist in Gothenburg had just succeeded in detecting changes in the composition of nucleic acids in the brains of rats taught a wire-climbing trick. I called the radio station immediately for a copy of the story from the newswire; and when it arrived the next day, I wrote immediately to the scientist named in the story, Holger Hydén, for reprints of his research publications.

When the papers from Sweden arrived, they unveiled a fascinating probe into the chemistry of brain function. Hydén had succeeded in measuring changes in RNA concentration, and in the base composition of the RNA, in tiny amounts of brain tissue of rats trained to climb a narrow wire to reach a food reward. Hydén had taken tissue samples from the vestibular nucleus, the region of the brain involved in the difficult motor learning task. These changes in RNA meant that the chemical composition of the nerve cells of the vestibular nucleus was changed by the animal's experience in learning to climb the wire.

Though new to me, this was not the first time Hydén had shown that cells in the brain show chemical changes in response to stimulation. In 1943, he received a medical degree from the Karolinska Institute in Stockholm, the city of his birth in 1917, and published a paper using elegant microanalytical techniques to show that the protein content of individual nerve cells could vary with their state of stimulation.[1] In 1950, he reported that the protein content of cells in the brain stem area concerned with posture and balance was increased in rabbits subjected to prolonged rotation.[2] Following his graduation from medical school, he continued doing research at the Karolinska Institute until being appointed as a professor of histology at the University of Gothenburg in 1949.

Hydén's research productivity was interrupted during the mid-1950s by several years of administrative duties (culminating in the presidency of the university in 1957), but the prospect of heading up a new institute of neurobiology tempted him back into the lab by the end of the decade. As the 1960s began, three trademarks of Hydén's career were evident. First, he had become the world's leader in freehand microdissection of individual brain cells and the analysis of their macromolecular composition. Secondly, he began to insist on the importance of neuroglia—the non-neuronal cells intimately entwined with neurons—as part of a neuron-glia functional unit. And thirdly, he began to offer specific hypotheses about how macromolecules could encode behavioral information, through a sequence of publications that showed a conceptual evolution in sophistication that tracked newly acquired molecular biological and neurochemical information as it became available.

Hydén's discovery of base ratio changes in the nucleotides of RNA in the vestibular nucleus of rats taught to balance on a wire in 1962[3] was followed by a similar finding in cortical cells of rats forced to change their hand preference in reaching for food in 1964.[4] These results generated a fairly simplistic hypothesis: (1) The altered frequency of excitation of a neuron newly involved in a learning task disrupts ionic balance within the neuron, weakening stability of one or more bases in RNA, which are replaced by different bases from the available pool, leading to a permanent alteration in the cell's RNA. (2) The new RNA codes for a slightly different protein that activates more effectively a neurotransmitter. (3) The altered effectiveness of synaptic transmission represents a degree of synaptic plasticity. (4) Recurrence of the same pattern of excitation subsequently activates the same synaptic connection more effectively, reflecting a permanent form of neural plasticity.[5]

In 1969, Hydén reported the results of a collaboration with E. Roy John, showing that base ratio changes occurred in planaria classically conditioned to respond to a light in anticipation of an electric shock. However, base ratio changes also occurred in the RNA of pseudoconditioned planaria, in which the light and electric shock were randomly paired.[6] In any event, the overall experience led to changes in RNA, whether or not specifically related to learning.[7] The next iteration of Hydén's thinking about the mechanism of neural plasticity showed evidence of John's influence, as well as compatibility with some of Ross Adey's concepts: (1) New behavioral associations lead to diffuse changes in ionic equilibration over dispersed fields in the brain. (2) Changes in the pattern of ionic fluxes in the affected population of brain cells induce conformational (shape) changes in certain proteins that derepress segments of the genome, resulting in transcription of novel RNA segments. (3) This new RNA gets translated to new protein, which possibly gets incorporated permanently into that neuron's membrane. (4) Over time, each neuron acquires its unique pattern of specific proteins, which determine whether it responds to a particular pattern of ionic fields. (6) During learning, similar RNA is produced in glia, which may migrate from glia to neurons to maintain the genomic derepression.[8] Thus, glia and neurons become functionally coupled. Robert Galambos had earlier proposed a functional coupling between neurons and glia on theoretical grounds, but without molecular details or the data that Hydén's experiments provided.[9]

Hydén's legacy in some ways is less than he deserved because of what was arguably his greatest talent: a facility for microdissection of single cells from the brain and the ability to analyze their nucleic acid and protein composition. In his choice of complex motor learning tasks, he also blazed a trail that almost no one else tried to follow. He was skilled and creative in both his cellular and his behavioral manipulations, but in their detail, they stand apart from the mainstream of research on the biochemical basis of learning. Nevertheless, he was almost certainly correct in proposing that learning involves the transcription of new RNA, resulting in the synthesis of either novel proteins or more of a preexisting protein, and he was the first to show in a compelling way that new behavioral experiences have biochemical consequences at the cellular level in the brain.

Mark Rosenzweig, David Krech, and Edward Bennett

In the summer of 1947, Donald Hebb taught a graduate seminar as a visiting professor at Harvard. One of his students was Mark Rosenzweig, who was working with Walter Rosenblith and Robert Galambos on the neurophysiology of hearing. Rosenzweig was born and raised in Rochester, New York, earning a bachelor's degree and starting to graduate school at the University of Rochester when World War II intervened. He joined the navy, hoping to get a research appointment to the naval hospital in Bethesda, but his expertise in electronics earned him an assignment as a radar technician on a seaplane tender in China instead. After the war, he returned to graduate school, but at Harvard this time.

In 1951, Rosenzweig was hired as an assistant professor of psychology at the University of California in Berkeley, where he would spend his entire career. There he me David Krech, another psychologist who had spent years analyzing maze learning in rats. Krech was convinced that different rats use different learning strategies and began to wonder if different parts of the brain—say the visual cortex for visual learners—would show biochemical changes more than other regions of the brain. Rosenzweig was intrigued with this idea, but they needed a biochemist if they were going to take up investigation of chemical changes in the brain. One of the biochemists recommended to them was Ed Bennett, who had been recruited to Berkeley by Melvin Calvin, a Nobel laureate for his work on photosynthesis.

Bennett was dubious, but initial results indicated that there were indeed regional differences in cholinesterase,[10] the enzyme that breaks down the neurotransmitter acetylcholine (ACh), which would alter the efficacy of synaptic function. The more they thought about it though, the more they began to think that the changes might be associated with the overall complexity of the task rather than any particular learning strategy. Rosenzweig then thought back to what he had heard Hebb describe in the graduate seminar at Harvard about taking a few rats home as pets. After weeks of free reign at his house, Hebb seemed to think that the pet rats behaved more intelligently than the lab rats left in their home cages, unstimulated by the rich variety of sensory experiences that free exploration of a human household afforded. They decided, therefore, to launch an investigation of the effects of an enriched environment on the brain.

Rosenzweig, Krech, and Bennett constructed elaborate playpens with toys, mobiles, objects for climbing and exploration, and other devices intended to provide a more stimulating environment filled with opportunities for learning. They called this the environmental complexity and training (ECT) condition. Contrary to their expectations that cholinesterase would be increased in the ECT rats, they found the opposite. On closer examination of their data, however, it appeared that the weight of the cortex, which was used to normalize the samples, was increasing more than the cholinesterase activity was changing, so the ratio of enzyme activity to tissue weight was going down.[11] Therefore they decided they needed an anatomist.

Marion Diamond was recruited for this task. Her careful histological measurements revealed that there was indeed a small but statistically significant increase in cortical thickness in the ECT rats.[12] By 1964, it was known that the brain has a specific form of cholinesterase called acetylcholinesterase (AChE), so the Berkeley group starting reporting their results on this more brain-specific form of the enzyme, confirming small but significant increases of it in the ECT rats.[13] Subsequent studies would add morphological detail, such as an increase in glia[14] and morphological changes at the synapse.[15]

The findings of the research at Berkeley was met with a lot of initial skepticism because the changes were so small and biochemists at the time were not used to evaluating statistical data. Also, the results went against the prevailing assumption that the mature brain simply doesn't change morphologically and that biochemical pathways are genetically fixed. Even after the results of Rosenzweig and his colleagues were generally accepted, they explained little about the mechanisms of learning and memory, since the ECT paradigm was behaviorally so complex and therefore activated such widespread brain activity. But what their research left no doubt about was the fact that the brain is structurally plastic and susceptible to long-term metabolic alteration. Hydén had shown that genomic readout into RNA, hence presumably into new protein, was a consequence of learning. Rosenzweig, Krech, Bennett, and Diamond showed ample ways in which new gene expression could be part of the way that memory must be encoded in a plastic brain. The next step was to pin down a mechanism for memory encoding in a nervous system somewhere in the animal kingdom. This required a level of simplification and reduction that few thought could be successful.

Eric Kandel

Eric Kandel was born to Jewish parents in Vienna in 1929, barely reaching an age old enough to escape the Anchluss of Germany and Austria by immigrating to the United States with his brother in 1939. He had been appalled by the way the cultured population of Vienna had turned on its Jewish neighbors when allowed to do so by the new Nazi government.[16] Perhaps that close call with the darker side of humanity, followed by the liberating feeling of his new home in Brooklyn, gave him a lifelong tendency to treasure his own free choices and follow his intuitions.

He was one of two students out of over a thousand in his high school graduating class to be admitted to Harvard College. Enthralled with the Freudian psychoanalysis prevalent at the time, he decided to go to medical school in order to become a psychiatrist. He was admitted to the New York University Medical School in 1952 and graduated in 1956, but during his final year, he did an elective research rotation with Harry Grundfest at Columbia University. Grundfest listened patiently to Kandel's enthusiasm about searching for the Freudian ego and id in the architecture of the brain, then explained that such a quest was well beyond the state of technology at the time. If we want to understand mind, Grundfest claimed, we have to look at the brain "one cell at a time."[17] Though initially demoralized by this reality check, Kandel in time would turn the advice into an overwhelmingly successful strategy.

During his brief tenure in Grundfest's lab, Kandel learned from Dominick Purpura how to design and carry out electophysiological experiments on the brain. He soon realized, however, that the mammalian brain was challenging for the study of unit recordings of "one cell at a time." This coincided with the publication of studies on crayfish neurons by Stephen Kuffler, which inspired Kandel to look more favorably toward the simpler nervous systems of invertebrates, with their smaller number and larger size of neurons. During his year of internship at Montefiore Hospital in New York, he returned for a second period of brief work with Grundfest, this time to work with Stanley Crain on recording the electrical activity of nerve cells in the crayfish. Of this episode, Kandel reflected, "I was becoming a true psychoanalyst. I was listening in to the deep, hidden thoughts of my crayfish."[18]

On the strength of these two periods in his lab, Grundfest nominated Kandel for an appointment to the National Institute of Mental Health (NIMH). Soon after his arrival, Kandel read about Brenda Milner's study of a patient who lost the ability to retain short-term memory following surgical ablation of regions of his cerebrum that included the hippocampus.[19] Since the patient's long-term memory and ability to function normally were unaffected, Milner concluded that short-term memory must reside in the hippocampus. Kandel decided he wanted to pick up where Milner had left off—to study the neurons of the hippocampus to see if they differed in some electrophysiological way from spinal motor neurons, which presumably were not involved in memory and were the only other nerve cells in mammals studied in detail to that point. He and Alden Spencer, another postdoc in the same lab, were able to record the first single-cell activity from the hippocampus, which gave his ego and confidence a boost. One of the first things they discovered is that some neurons in the hippocampus fire spontaneously and seemingly at random—a finding that would have pleased Roy John. But otherwise, their results showed no particular difference from the characteristics of spinal motoneurons, hence no functional difference between memory-encoding cells of the hippocampus and motor neurons in the spinal cord that presumably played no role in memory formation. On the basis of those results, Kandel and Spencer concluded that "the cellular mechanisms of learning and memory reside not in the special properties of the neuron itself, but in the connections it receives and makes with other cells."[20] But how are those connections increased, strengthened, or weakened?

As the period for his appointment at NIMH approached an end, Kandel returned to his conviction that a simpler system was needed for studying the changes that occur with memory deposition. After several months of contemplation, he settled on the giant marine snail, *Aplysia*, which he learned from a lecture by Angelique Arvanitaki-Chalazonitis is a large mollusk—over a foot long—with a relatively small total number of neurons—about twenty thousand. Importantly, all the neurons are uniquely identifiable, and some are quite large, making location of the same cell in the network reliable from one animal to the next, and penetration by electrodes easy. The only other person working on *Aplysia* at the time was a French biophysicist, Ladislav Tauc. In early 1960, Kandel visited Tauc and arranged to go to Paris to work with him at the end of his residency in 1962.

The demands of Kandel's residency at the Massachusetts Mental Health Center of Harvard Medical School were light, and this, plus a falling out with his wife over her complaint that he was working too hard and ignoring his family, led him to spend more time at home and more time thinking about what he would do with *Aplysia*. He decided that he would try to mimic the simplest forms of learning at the cellular level. He would mimic habituation by gently but repetitively stimulating a presynaptic neuron to see if the postsynaptic neuron would show a diminished response over time. He would mimic sensitization by zapping a separate neural pathway strongly, then testing to see if the presynaptic neuron would cause the postsynaptic cell to show a heightened response subsequently. And he would mimic classical conditioning by gently stimulating the presynaptic neuron, then zapping the other pathway more strongly, to see if subsequent low-level stimulation from the presynaptic neuron would cause a stronger response in the postsynaptic cell. These experimental designs were spelled out in an NIH postdoctoral proposal, which was awarded to Kandel in 1962, allowing him to head to Paris for fourteen months of work in Tauc's lab.

As expected of European emigres and connoisseurs of culture like Kandel and his wife, the experience of living in Paris was a delight.

More significantly for science, the work went very well in the lab. Each of Kandel's hypotheses was born out: habituation caused a diminished electrophysiological response in the postsynaptic cell, sensitization led to an amplified response, and conditioning likewise led to a greater response in the postsynaptic cell than had been elicited prior to the unconditioned shock in the alternate pathway. All these experiments were conducted on a ganglion (collection of nerve cells) removed from the animal and therefore could not be tied to behavior. But they showed that function at the synaptic level was plastic—under appropriate conditions, the strength of a postsynaptic response could be altered in a way that lasted for a period of time. Something analogous to learning had happened at a functional connection between nerve cells. The relevance of this finding to behavior was still a long way from demonstration, but it was most definitely a *thing* in a *place*.

[1]   Hydén, H. 1943. Protein metabolism in the nerve cell during growth and function. *Acta Physiol Scand* 6: 1–136.

[2]   Hydén, H. 1950. Spectroscopic studies on nerve cells in development, growth, and function. In *Genetic Neurology*, edited by P. Weiss. Chicago: Univ. of Chicago Press.

[3]   Hydén, H. and E. Egyházi. 1962. Nuclear RNA changes of nerve cells during a learning experiment in rats. *Proc Natl Acad Sci USA* 48: 1366–73.

[4]   Hydén, H. and E. Egyházi. 1964. Changes in RNA content and base composition in cortical neurons of rats in a learning experiment involving transfer of handedness. *Proc Natl Acad Sci USA* 52: 1030–5.

[5]   Hydén, H., ed. 1959. *Biochemical changes in glial cells and nerve cells*. Edited by F. Brücke. Vol. 3, pp. 66–89, *Fourth Intl. Congress of Biochemistry*.

[6]   While the intent of the experimental design was to isolate the effect of learning per se, as in all pseudoconditioning experiments, the fact that the organism probably learned something, even if not what the experimenter intended, cannot be excluded.

[7]   Hydén, H., E. Egyházi, E. R. John, and F. Bartlett. 1969. RNA base ratio changes in Planaria during conditioning. *J Neurochem* 16: 813–21.

[8]   Hydén, H. 1970. The question of a molecular basis for the memory trace. In Biology of Memory, edited by K. H. Pribram and D. E. Broadbent. New York: Academic Press.

[9]   Galambos, R. 1961. A glia-neural theory of brain function. *Proc Natl Acad Sci*

USA 47: 129–136.

[10] Krech, D., M. R. Rosenzweig, E. L. Bennett, and B. Krueckel. 1954. Enzyme concentrations in the brain and adjustive behavior-patterns. *Science* 120: 994–6; Krech, D., M. R. Rosenzweig, and E. L. Bennett. 1959. Correlation between brain cholinesterase and brain weight within two strains of rats. *Am J Physiol* 196: 31–32.

[11] Krech, D., M. R. Rosenzweig, and E. L. Bennett. 1960. Effects of environmental complexity and training on brain chemistry. *J Comp Physiol Psychol* 53: 509–19.

[12] Rosenzweig, M. R., D. Krech, E. L. Bennett, and M. C. Diamond. 1962. Effects of environmental complexity and training on brain chemistry and anatomy: a replication and extension. *J Comp Physiol Psychol* 55: 429–37.

[13] Rosenzweig, M. R., E. L. Bennett, and D. Krech. 1964. Cerebral Effects of Environmental Complexity and Training among Adult Rats. *J Comp Physiol Psychol* 57: 438–9; Bennett, E. L., M. C. Diamond, D. Krech, and M. R. Rosenzweig. 1964. Chemical and anatomical plasticity of the brain. *Science* 146: 610–619.

[14] Diamond, M. C., F. Law, H. Rhodes, B. Lindner, et al. 1966. Increases in cortical depth and glia numbers in rats subjected to enriched environment. *J Comp Neurol* 128: 117–26.

[15] Diamond, M. C., B. Lindner, R. Johnson, E. L. Bennett, and M. R. Rosenzweig. 1975. Differences in occipital cortical synapses from environmentally enriched, impoverished, and standard colony rats. *J Neurosci Res* 1: 109–19.

[16] Kandel, E. R. 2006. *In Search of Memory: The Emergence of a New Science of Mind*. New York: W. W. Norton.

[17] Kandel, E. R. 2006. *op. cit.,* p. 55

[18] Kandel, E. R. 2006. *op. cit.,* p. 109

[19] Scoville, W. B. and B. Milner. 1957. Loss of recent memory after bilateral hippocampal lesion. *J Neurol Neurosurg Psychiat.* 20: 411–421.

[20] Kandel, E. R. 2006. *op. cit.,* p. 142

6

Disrupting Memory

While the efforts during the early 1960s of Hydén, Rosenzweig, Krech, Bennett, and Kandel were concentrated on establishing that memory deposition requires the elaboration of specific molecules, or structural changes in the brain, or electrophysiological adaptations at the synapse, another strategy was also being tested: interference with or enhancement of the brain's ability to encode memory. Several investigators reasoned that by disrupting or enhancing the deposition of memory, they might gain insight into the mechanisms involved. The two most common approaches were (1) an attempt to block macromolecular synthesis, which would impact encoding through molecular means at a specific site, and (2) electrocortical shock, which would affect dispersed processes in a population.

Louis Flexner

Louis Barkhouse Flexner was born in 1902 in Louisville, Kentucky, into a distinguished medical family. One uncle, Simon Flexner, was a professor of pathology at the University of Pennsylvania Medical School before becoming president of the Rockefeller Institute for Medical Research (later, Rockefeller University). Another uncle, Abraham Flexner, became an organizer and first director of the Institute for

Advanced Study at Princeton University. Louis received a bachelor of science degree from the University of Chicago in 1923, and following the family tradition, studied for a medical degree at Johns Hopkins, which he completed with difficulty in 1927. He found the pedantry and chaos of medical education at the time ill-suited to his educational philosophy and managed to finish at the bottom of his graduating class (along with an eventual Nobel laureate, Keffer Hartline).[1]

Notwithstanding this inauspicious beginning, he found biochemistry attractive enough to stay on at Hopkins for three more year, working with Leonor Michaelis on redox potentials in biochemical reactions. A year of interning at the University of Chicago clinics was less satisfactory for him than working in a research laboratory, so he returned to Johns Hopkins for nine years of research, primarily on cerebrospinal fluid. Early in this phase of his career, he met a dynamic woman from Catalonia who had earned a doctorate in pharmacy from the University of Madrid in 1927. Josefa was driven from Spain by the impending civil war and ended up at Johns Hopkins where she and Louis met and formed a partnership in marriage and science that lasted over sixty years.

After their marriage in 1938, the Flexners moved to the Carnegie Institute in 1941, where they worked together primarily on embryological development. In 1951, Louis was recruited to be chair of anatomy at the University of Pennsylvania Medical School, where he hoped to develop a faculty working at the forefront of knowledge, especially on (*a*) cell differentiation and (*b*) the structure and function of the central nervous system. In 1953, he organized the Institute for Neurological Sciences, which became of forerunner of numerous entities like it that would eventually be known as neuroscience departments or divisions.

In 1960, Louis, Josefa, and their colleagues announced the discovery that lactic dehydrogenase (LDH), a key enzyme in energy metabolic pathways, consists of four different forms, which vary in their proportional makeup of the overall enzyme from one tissue to the next and which shift among the forms as development proceeds.[2] This discovery, made independently also by Clement Markert and Freddy Moller,[3] added to the growing reputation of the Flexner lab and led to the election of Louis in 1964 to the National Academy of Sciences. Though the product of a distinguished family and a scientist at a succession of elite institutions, he was wary of elitism and sometimes declined to attend meetings of the National Academy because it had refused to admit some worthy scientists that he thought had been unjustly excluded.

Fresh from his discovery of four different catalytic variants (isozymes) of LDH, Flexner turned his attention to the biochemical basis of memory. To test the hypothesis that protein synthesis is necessary for encoding memory, he used the antibiotic puromycin in the hope of blocking protein turnover in the brain. He found that he could block incorporation of the amino acid valine into protein (a measure of protein synthesis) by up to 83 percent by injecting the antibiotic directly into the brain. But blocking protein synthesis by this much did not prevent mice from learning a conditioned avoidance task one to seven hours postinjection, nor was retention of Y-maze learning affected by puromycin injected two to eight hours after training.[4] In subsequent experiments, though, Flexner found that he could block retrieval of learning to avoid a shock by choosing the correct arm of a Y-maze when puromycin was injected into the temporal brain region, which included the hippocampus, within three days of training. If the injections were delayed longer, injections needed to be made into other cortical regions as well. Flexner interpreted these results to mean that the site of memory deposition spread from a localized area in or encompassing the hippocampus to a wider distribution of higher brain regions.

Flexner was aware that the case for protein synthesis involvement in learning was not airtight. The degree to which protein had to be suppressed was severe, and another antibiotic also known to inhibit protein synthesis (chloramphenicol) had no effect on memory.[5] He noted that the loss of memory might have resulted from effects other than protein synthesis inhibition—an admission that turned out to be prophetic. Also, the experimental design that Flexner used clearly was testing retrieval of the memory, not its deposition. So while the initial efforts to disrupt memory by the Flexner lab implicated a role for protein synthesis in memory, the experiments were far from convincing. They got a lot of attention though—in part because theirs was among the first attempts to confirm a biochemical basis for learning and, secondly, because of the rapid and exciting progress being made in molecular biology at the time.

Sam Barondes, as we have seen, was among those involved in the race to decipher the genetic code. Two of his colleagues at NIH, Wesley Dingman and Michael Sporn, were actually ahead of Flexner in using the approach of blocking macromolecular synthesis as a means of blocking memory, except that their target was RNA instead of protein. They had shown, just as Barondes was joining Nirenberg's lab, that injection of 8-azaguanine, an analog of an RNA precursor that produces aberrant RNA, slowed the ability of rats to learn how to escape from a water maze when the drug was injected into the ventricles of the brain.[6] Barondes decided to repeat their approach using a different antibiotic, actinomycin D. He found that inhibition of RNA synthesis by up to 86 percent did not prevent rats from learning to avoid a shock, despite drastic side effects from the drug.[7] In fact, the toxicity of actinomycin D was so great that Barondes retreated to the use of puromycin as Flexner had done. He injected puromycin into both sides of the temporal region of the brain of rats five hours before training them to avoid shock by choosing the correct limb of a Y-maze. When retested fifteen minutes after training, the rats had normal retention. But over the ensuing hours, the puromycin-injected rats, unlike the controls injected with saline, showed a progressive decrease in remembering how to avoid the shock (down to 91 percent loss of memory three hours after training). Combining these results with Flexner's, Barondes and Cohen concluded that memory becomes fixed in three phases: initially, through a process insensitive to puromycin; then later through a process localized in the temporal lobe (encompassing the hippocampus), and later still through a process more broadly spread throughout the brain.[8]

Bernard Agranoff

Bernard "Bernie" Agranoff was born in 1926 and raised in Detroit, where he attended Cass Tech, one of the nation's leading magnet schools at the time. His initial interest was primarily in art, but when his art teacher on the first day of class justified her assertion that two triangles are similar if they share two sides and an angle in common because "you feel it in your bones," Agranoff decided that maybe art wasn't for him. He next tried architecture, then drifted toward science just as World War II was beginning. Despite lacking a foreign language prerequisite, he was admitted into the Naval Officers Training Program at the University of Michigan. This provided a fast track to medical school (one of five options in the program), which appealed to his interest in science. When he discovered he could enroll in the medical school at Wayne State University in Detroit a year earlier than in the University of Michigan's medical school, he opted for the former.[9]

Agranoff didn't care for medical school and almost quit midway through the curriculum, but his professors talked him out of it, and he received his medical degree in 1950. Gordon Scott, the dean of the medical school at Wayne State, recognized Agranoff's talent for research and recommended him to Francis O. Schmitt, a biophysicist (before the term had been invented) at the Massachusetts Institute of Technology (MIT). Schmitt talked him into taking the preliminary exam for a PhD, which he passed, but before he could finish his research for that degree, the Korean War began, and he was taken in the physician's draft by the navy for assignment to the naval hospital in Bethesda, Maryland. There, he met Roscoe Brady, who had set up a neurochemistry lab at the NIH across the street. When his two-year military obligation was completed, Agranoff was offered a chance to return to Schmitt's lab at MIT, but Brady offered him a chance at independence, which appealed to him more. So in the fall of 1954, Agranoff began a productive period of research in Roscoe Brady's unit at NIH.

As his appointment at NIH was beginning, his interest in the biochemistry of brain function was increasing. Chlorpromazine (Thorazine) had been introduced as the first antipsychotic drug in

1950, and meprobamate (Miltown) soon followed. Agranoff was drawn to psychopharmacology in part because his mother suffered from severe depression and his sister was a psychiatric social worker. He first turned his attention, as a biochemist, to the way meprobamate is metabolized by the body, using himself as a sample size of one. At the same time, his interest in behavior led him to try out the effect of meprobamate on imprinting in a group of ducklings that followed him around the halls of NIH after they were hatched in his presence. Ultimately, he decided that meprobamate was not going to give him any insight into the mechanism if imprinting, but it did start him thinking seriously about brain biochemistry and memory.

Meanwhile, he stuck mainly to his day job as a biochemist. His first major insight was to propose a novel CDP-liponucleotide (CDP is cytidine diphosphate, containing the same base that occurs in DNA and RNA), which turned out to be a key intermediate in the synthesis of a very important family of lipid compounds in the brain. Within a few months of that publication, he discovered a key metabolic step in the synthesis of cholesterol while working briefly in the Munich lab of Feodor Lynen, who would win the Nobel Prize in 1964 for his work on cholesterol metabolism. By the time Agranoff returned to NIH in 1958, his qualifications as a biochemist had thus been solidified, and he decided it was time to move on. An opening at his alma mater, the University of Michigan, provided a highly attractive opportunity in 1960.

Agranoff was drawn to the position in Ann Arbor especially because it provided a dual appointment in the Department of Biological Chemistry, where he could maintain his status as a card-carrying biochemist, and at the Mental Health Research Institute, where he could pursue his more speculative studies of the biochemical basis of behavior. His new lab was in the latter facility, where he met Jim McConnell (more about him later), a psychologist who had followed the same strategy as Kandel by pursuing the study of learning in a simple invertebrate. McConnell's choice was the planarian, a simple flatworm that has rudimentary ganglia for a "brain" in its head region and bilateral nerve cords with ladderlike interconnections (commissures) running the length of its body. His behavioral paradigm was classical conditioning to light followed by an electrical shock. Several pairings of light and shock, according to McConnell, led to reflex contortion of the planarian's small body upon presentation of the light alone.

Agranoff was skeptical of the subjective judgments of the worm's behavior that McConnell and his student assistants were collecting, but he was attracted by the possibility that these simple animals could prove adequate for correlating biochemistry with their behavior. He hired Paul Klinger as a lab assistant to set up an instrumental way of objectively measuring the behavioral responses of planaria but, after considerable effort, was unable to reliably detect the conditioned learning that McConnell had claimed. In frustration, he turned to guppies, noting with relief at the time that "they're loaded—they've got eyeballs and a spinal cord."[10] They also showed more robust behavior that was easy to objectively quantify. However, when a freeze in Florida wiped out his reliable supply of guppies, he turned to the larger goldfish instead.

The switch to goldfish was not arbitrary. His attention to them had been drawn dramatically by a visit to Jeff Bitterman's lab at Bryn Mawr College in Philadelphia, when Bitterman complained that his goldfish had robust long-term memory that interfered with his experiments on short-term learning. This was music to Agranoff's ears. The visit to Bitterman's lab came the same day as an earlier visit with Flexner at the University of Pennsylvania. By then, Agranoff knew of Flexner's attempt to interrupt memory encoding through the use of antibiotics, so he returned to Ann Arbor determined to see if the same would be true of animals with a real brain and spinal cord who apparently had an excellent memory.

Agranoff and Klinger trained the goldfish to avoid an electric shock by swimming across a barrier into a dark compartment of their aquarium. They found upon testing four days later that the fish remembered how to escape the shock if they had been injected intracranially with saline right after training, but not if they had been injected with puromycin.[11] Subsequent experiments showed that puromycin did not interfere with learning, but only with retention. He was also able to show that varying the time of puromycin injection after the end of training trials revealed that the longer the interval, the less the memory was disrupted.[12] In other words, puromycin didn't block acquisition, nor did it block retrieval if injected more than sixty minutes after the end of training, but it did block memory if injected right after training. Flexner had shown that memory in rodents was susceptible to inhibition for much longer than that, so interpretation of the results was murky. Perhaps goldfish and rats responded very differently to puromycin. Regardless of the details though, the picture that memory goes through different phases of consolidation was beginning to emerge.

Retrograde Amnesia

These early experiments with metabolic inhibitors seldom provided confirmatory evidence for or against a critical role for macromolecular synthesis in the storage of memory. The behavioral paradigms, species used, and time courses studied were too different for the construction of a coherent conclusion. Taken together, though, they were indicating that either consolidation or retrieval of the memory, or both, became more likely in progressive stages following the experience. Adding to this perception was a related class of experiments that caused retrograde amnesia, or the loss of ability to encode the memory due to interference with brain function after behavioral training.

Murray Jarvik, who had been one of Barondes' professors and a coauthor of his first experiment with puromycin, had earlier used anesthetics (ether and pentobarbital) on rats taught to avoid a minor electric shock. He found that ether administered within five minutes, or pentobarbital injected within ten minutes, of the training session prevented rats from remembering how to avoid the shock the next day. If the rats were anesthetized hours after their training, however, they remembered the task the following day.[13] Jarvik and his coauthors concluded that the anesthetics were interfering with a fixation process (consolidation) that was time dependent.

Another approach was the use of electroconvulsive shock to cause retrograde amnesia. One of the first scientists to try this approach was

James McGaugh, who had worked with Rosenzweig and Krech for his doctorate. McGaugh was born and grew up in Southern California, attending San Jose State as a music and drama major until he saw a picture of a neuron for the first time in his introductory psychology course and thought, "If we knew how that cell worked, my memorization in music and drama would be a lot easier."[14] With his doctorate from Berkeley in hand, he proceeded to develop a learning situation in which rats, which have a natural tendency to step down from a pedestal, learned not to do so in order to avoid an electric shock. This was essentially a one-trial learning task, so the time from the experience that would be fixed into memory could be known with precision. McGaugh found that an electrocortical shock (ECS) given at longer intervals between the training and ECS were increasingly less effective at preventing the rats from remembering not to step down from the platform. Thus, ECS appeared to interfere with consolidation of the memory of how to avoid the shock, but in a manner that diminished with time.[15]

One of the problems with McGaugh's early experiments was that the ECS he used was so strong that it caused convulsions, so it was impossible to tell if the retroactive amnesia was due to electrical interference with the brain's normal activity or the traumatizing effect of the convulsions. So McGaugh then anesthetized his rats prior to administering the ECS to block the convulsions and found that the ECS was still effective in causing amnesia if administered soon enough after the one-trial learning experience.[16] It was the electrical chaos in the brain, not the convulsions themselves, responsible for the amnesia.

Agranoff also used ECS as a way to disrupt memory in his goldfish. Just as he had done with puromycin, he varied the interval between training the fish to escape from shock and administration of ECS. The curves for retention of memory improved as the interval between training and ECS administration increased.[12]

In 1965, both Barondes and Agranoff published summaries of the state of biochemical studies of memory, and McGaugh did so the next year. All three authors concluded that evidence was ample by the mid-1960s that brain mechanisms of memory encoding consisted of graded phases, in which the earliest stage—a form of short-term memory

(STM)—is labile to disruption, while the later stage, or long-term memory (LTM), is more durable. Agranoff wrote,

> [I]t would seem that short-term or temporary memory involves electrical states, conformational changes in protein, or other readily reversible phenomena, while long-term or permanent memory formation requires metabolic changes of which protein synthesis is a part.[17]

McGaugh suggested that there might be three memory trace systems: "one immediate memory, one for short-term memory . . . and one which consolidates slowly and is rather permanent."[18]

Barondes was explicit in proposing "that memory is based on the formation of new synaptic connections, and that proteins are involved in the process." However, more clearly than anyone else at the time, he made the distinction between proteins as an *embodiment* of memory (equivalent somehow to a *thing in a place* that constitutes the engram, or storage form of the memory), as opposed to being part of a structural or functional change occurring in a neural matrix, where the matrix itself constitutes the engram. While not stating so himself, this was a role for molecules more compatible with the notion of a *process in a population.*[19]

[1]   Sprague, J. M. 1998. Louis Barkhouse Flexner (1902-1996). In *Biographical Memoirs*. Washington, DC: Natl. Acad. Sci. USA.

[2]   Flexner, L. B., J. B. Flexner, R. B. Roberts, and G. De La Haba. 1960. Lactic dehydrogenases of the developing cerebral cortex and liver of the mouse and guinea pig. *Dev Biol* 2: 313–28.

[3]   Markert, C. L. and F. Moller. 1959. Multiple forms of enzymes: tissue, ontogenetic, and species specific patterns. *Proc Natl Acad Sci USA* 45: 753–63.

[4]   Flexner, J. B., L. B. Flexner, E. Stellar, G. De La Haba, and R. B. Roberts. 1962. Inhibition of protein synthesis in brain and learning and memory following puromycin. *J Neurochem* 9: 595–605.

[5]   Flexner, L. B., J. B. Flexner, R. B. Roberts, and G. DelaHaba. 1964. Loss of recent memory in mice as related to regional inhibition of cerebral protein synthesis. *Proc Natl Acad Sci USA* 52: 1165–9.

[6]   Dingman, W. and M. B. Sporn. 1961. The incorporation of 8-azaguanine into

rat brain RNA and its effect on maze-learning by the rat: an inquiry into the biochemical basis of memory. *J Psychiatr Res* 1: 1–11.

[7]   Barondes, S. H. and M. E. Jarvik. 1964. The influence of actinomycin D on brain RNA synthesis and on memory. *J Neurochem* 11: 187–95.

[8]   Barondes, S. H. and H. D. Cohen. 1966. Puromycin effect on successive phases of memory storage. *Science* 151: 594–5.

[9]   Agranoff, B. W. 2009. Bernard W. Agranoff. In *The History of Neuroscience in Autobiography*, edited by L. Squire. Washington, DC: Society for Neuroscience.

[10] Irwin, L. N. 2007. *Scotophobin: Darkness at the Dawn of the Search for Memory Molecules*. Lanham, MD: Hamilton Books. Further details of Agranoff's career and behavioral research are also found in that work.

[11] Agranoff, B. W. and P. D. Klinger. 1964. Puromycin effect on memory fixation in the goldfish. *Science* 146: 952–3.

[12] Davis, R. E., P. J. Bright, and B. W. Agranoff. 1965. Effect of ECS and puromycin on memory in fish. *J Comp Physiol Psychol* 60: 162–6.

[13] Pearlman, C., S. Sharpless, and M. E. Jarvik. 1961. Retrograde amnesia by anesthetic and convulsant agents. *J Comp Physiol Psychol* 54: 109–112.

[14] Irwin, L. N. 2007. *Scotophobin: Darkness at the Dawn of the Search for Memory Molecules*. Lanham, MD: Hamilton, p. 89. More information on McGaugh is given in this book.

[15] Hudspeth, W. J., J. L. McGaugh, and C. W. Thomson. 1964. Aversive and amnesic effects of electroconvulsive shock. *J Comp Physiol Psychol* 57: 61–4.

[16] McGaugh, J. L. and H. P. Alpern. 1966. Effects of electroshock on memory: amnesia without convulsions. *Science* 152: 665–6.

[17] Agranoff, B. W. 1965. Molecules and memories. *Perspect Biol Med* 9: 13–22.

[18] McGaugh, J. L. 1966. Time-dependent processes in memory storage. *Science* 153: 1351–8.

[19] Barondes, S. H. 1965. Relationship of biological regulatory mechanisms to learning and memory. *Nature* 205: 18–21.

7

Birth of the Neurosciences
Research Program

In the history of biophysics and neuroscience, there was no one quite like Frank Schmitt. To the generation of brain researchers coming of age during the 1960s and '70s, many (including myself) were often put off by his obsequious reference to status and authority, his incessant name-dropping, and his often naive speculations about brain function. But this was more a measure of our ignorance than a true reflection of him as a scientist. He conceived of and founded the Neurosciences Research Program (NRP), which, according to some, played a pivotal role in the development of neuroscience as a major interdisciplinary field of science in its own right and, in the view of others, was a peripheral player that simply rode the wave of enthusiasm for the newly integrating approaches to studying the brain. The ultimate impact of NRP, which will be dealt with as the rest of this book unfolds, is still not clear. But no history of the brain sciences can be considered complete without reflecting on its existence during the inaugural decades of the development of neuroscience.

Francis Otto Schmitt

Francis Otto Schmitt was born on November 23, 1903, in St. Louis, where he grew up and attended Washington University, earning an AB degree in 1924.[1] He proceeded to graduate school at the same institution, benefitting from the tutelage of Joseph Erlanger and Herbert Gasser, who were carrying out their famous experiments of nerve cell conduction at the time. He completed the work for the doctorate in three years, defending his thesis on conduction of excitation in cardiac muscle on May 21, 1927, the day that Charles Lindbergh landed in Paris. A month later, he married Barbara Hecker and took her on their honeymoon to the Woods Hole Marine Biological Laboratory where he hoped to spend time writing his thesis for publication. However, in his own words, "I don't think I ever put pencil to paper during that wonderful six weeks."[2]

There followed two brief years of postdoctoral study, one in chemistry at the University of California–Berkeley, and the second at University College in London. During this time, he began studies on the characteristics of lipids in membranes. From London, he moved to the Kaiser Wilhelm Institute in Berlin, where he had arranged to work with Otto Warburg, a pioneer in the study of cellular metabolism. The project he was assigned to by Warburg, however, was uninteresting, so he moved to a different project on the energetics of nerve cell conduction in the lab of Otto Meyerhoff.

German science was nearing its peak in those years, prior to the Nazi takeover, and Nobel prizes were being awarded to Germans and Austrians with regularity. Schmitt was deeply impressed by the quality of the science he witnessed, as well as the hierarchical meritocracy that he apparently felt accounted for its quality. He brought those impressions with him back to St. Louis, where he had arranged an appointment as assistant professor of zoology at his alma mater. For the next twelve years, he built a first-class research program in surface biochemistry and membrane ultrastructure, focusing especially on the orientation of membrane molecules in myelin, the lipid-rich wrapping around the axons of neurons. As his application of physical methods to the study of biological structures gained fame, so did his attractiveness to the Massachusetts Institute of Technology (MIT), which wanted to establish a bridge between engineering and biology. When they offered him the opportunity to form a new department in biology and biological engineering, he found the prospects of achieving his long-held vision of combining physics and biology too enticing to pass up. Thus he joined the faculty at MIT as a department head in 1941. His first goal was to set up a program using electron microscopy to study the ultrastructure of collagen. But soon, the war intervened, and his lab was diverted by the war effort to research on the best sutures for promoting wound healing, as well as research on rubber and other artificial elastomers.

After the war, he resumed his studies on collagen and developed one of the foremost electron microscopy labs in the world. Among those who came to MIT to avail themselves of his expertise and equipment were two British scientists, Jean Hanson and Hugh Huxley, who began the research that would lead years later to the sliding filament theory of muscle contraction. Through productive collaborations such as these, he continued to build a reputation for productivity that led to his election in 1948 to the National Academy of Sciences and induced MIT to make him an Institute Professor in 1951 (only the second one in its history), which freed him from administrative duties to pursue nothing but research full time.

A Biophysical Study Program

About the time Schmitt was made an Institute Professor at MIT, the grants division at NIH started getting more interdisciplinary proposals on physical approaches to biological problems and decided to set up a new study section on biophysics, which they asked Schmitt to chair. He was reluctant, not certain that he had the time or was up to the task. But a stronger feeling was his desire to promote a true merger between physics and biology, and he realized this would be a unique opportunity to bring that about. This decision foreshadowed one of his hallmark characteristics, especially of his later career. He was cautious and somewhat defensive about treading into new territory, but his ambition and resort to the authority of others often sufficed to compel him to take chances in bold directions.

It readily became apparent to the newly constituted study section that there was no consensus on what exactly constituted biophysics in 1956. Another of Schmitt's hallmarks was that once he had decided to take a bold plunge, nothing less than a big plunge would suffice. While a two-day conference of scientists actually working on biophysical problems might have satisfied many in the field, for Schmitt, it had to be a month-long meeting by a selection of the world's most eminent physicists and biologists. Such was his stature by then and his influence within the NIH and the National Academy, that he managed to get the funding for holding such a meeting, with travel expenses and housing for about 250 eminent scientists and their family members (since asking great men to desert their families for four weeks was unreasonable). The month-long "study program" on biophysics was held from July 20 to August 16 in 1958 in the pleasant surroundings of the University of Colorado campus in Boulder. The final product of the meeting was a volume entitled *Biophysical Science—A Study Program*, along with a clarified notion of what biophysics was about. For the participants, it was a great success; and for Schmitt, it was vindication of his belief that when you get the best and brightest minds from disparate fields together, intellectual hybrid vigor is bound to result.

The House that Schmitt Built

While Schmitt had, to this point, focused on a highly reductionistic approach to nerve function—what *he* was calling molecular biology before the term came to mean essentially molecular genetics—he actually harbored thoughts about higher brain function. One of his notions had been "the possibility that information might be transferred in the brain and central nervous system not only by electrical action waves along neuronal nets, but also by fast transport, *possibly through extracellular substances* [italics added]."[3] This idea would later find resonance with the work of Ross Adey, but at that time, it was purely unconventional speculation. Pursuit of the idea beyond the bounds of his Biophysical Study Program led him to organize a colloquium at MIT that resulted in a small volume[4] that did little more than restate the prevailing limited knowledge about the biological basis of memory at the time. Schmitt already had a bigger goal in mind.

Buoyed by the success of the Biophysical Study Program in 1958, Schmitt began to envision an ambitious follow-up on the same scale that would catalyze an interdisciplinary attack on the function of the brain, including what he called psychic biophysics. In his words, he wanted "to bring together leaders in many fields cognate to the study of the nervous system and the brain, and to establish a means by which meaningful interaction among these scientists might be facilitated."

His first task was to organize "a 'Core Group' of about ten to fifteen colleagues of eminence in the field." This group met for the first time on February 1, 1962, at a friend's apartment in Manhattan. Among those present of relevance to the history recounted here were Holger Hydén, Severo Ochoa, Robert Galambos, and Paul Weiss. He presented his grandiose plan and persuaded the assembled scientists of eminence to approve it unanimously. Within a month, NIH convened a special panel to consider a pilot grant to get the program underway.

Schmitt then set about to bring the NRP into existence through a series of actions. First, he would need to get the NRP authorized for sponsorship by MIT. Second, an inaugural group of Associates—the eminent scientists who would constitute the invisible college of the program—had to be chosen. Third, a private foundation to provide financial strength and program continuity—eventually named the Neurosciences Institute—had to be incorporated. Finally, a headquarters had to be identified and outfitted appropriately.

Given his stature at MIT, getting that institution to sponsor the NRP was a foregone conclusion. A group of eminent scientists was readily assembled as the inaugural class of Associates. Notwithstanding Schmitt's ultimate goal of understanding higher brain function, by the first anniversary of NRP's existence, it consisted of eight physicists and physical chemists, four chemists and biochemists, two molecular biologists, and two neurobiologists, but only two psychologists and one psychiatrist. The Neurosciences Research Foundation was established to pay for things like extravagant receptions, dinners, first-class airfare for Schmitt, and other expenses which government regulations would not allow. Through his acquaintance with Hudson Hoagland, then president of the American Academy of Arts and Sciences, Schmitt was able to procure a long-term lease of the third floor of the Academy's home at the Brandegee Estate—a palatial manor on a high hill in Brookline in the metropolitan Boston area. Renovations of those premises were completed by mid-October of 1962. By the end of that month, NIH had decided to award NRP not just a pilot program grant but a full five years of funding. NRP moved in and was up and running.

Among the first consequential hires at NRP was Ted Melnechuck as communications director. Arguably the most creative person Schmitt could have found, Melnechuck was highly talented and provocative, lending NRP a flair and level of daring that more conservative forces would rein in over time. Among Melnechuck's most consequential acts was initiation of the *NRP Bulletin*, an irregularly produced publication that started principally as a house organ, but turned into a vehicle for reporting the work sessions held on specific topics at NRP. At its peak, the *NRP Bulletin* had over 3,500 libraries and other subscribers and was widely seen as the most useful product of NRP available to the broader scientific community.

In 1964, Schmitt recruited Kay Cusick as business manager who served Schmitt and NRP for the entirety of its existence at the Brandegee Estate. He also hired that year as librarian for the NRP George Adelman, an English major who knew nothing about the brain and little about the rest of science, but who was swept up by the enthusiasm that Schmitt could engender and became one of NRP's most steadfast and devoted professional staff members. He also in time would become my closest friend when I eventually entered the NRP orbit for the first time four years later.

Work Sessions and Stated Meetings for "Interthinking"

The first institutional activity enacted by NRP on a recurring basis was the Stated Meetings of the Associates. Recall that the Associates were "of the highest scientific caliber." Having them gather together for two and a half days and discuss a preplanned topic usually of great concern to Schmitt was his fantasy. At Stated Meetings, "the air was usually charged with intellectual excitement," Schmitt wrote with satisfaction. "This was the atmosphere in which 'interthinking' occurred [through] highly informed input from experts in different fields."[5] There was no doubt that this was true, as I would observe upon attending them as a member of the staff a few years later. My doctoral adviser, who also served as a visiting scientist at NRP over a number of years, reminded me on more than one occasion that "it's a great privilege to sit in that room" where the Stated Meetings took place. What I wondered about, and what NIH eventually came to question, was whether a greater effort should be made to extend the benefit of NRP's existence to those in the scientific community *not* privileged to sit in that room.

NRP's response to that concern was to engage in a bit of window-dressing by bringing in visiting scientists, usually junior in status, for appointments of a few months to a couple of years. They were given the title of Staff Scientist and assigned a variety of tasks ranging from the trivial to the substantive, with outcomes that varied. Some were successful in attracting Schmitt's attention enough to take them into his confidence and even make them coauthors of his work on rare occasions. Many, however, were essentially ignored.

A more serious and successful way that NRP benefitted the scientific community was through work sessions, which also were two-and-a-half day meetings and likewise consisted of highly accomplished participants, but not necessarily Associates. The presentations at work sessions were written up by the chair with the help of NRP staff and published in the *NRP Bulletin* for wide dissemination. The weak link in this process was that writing up the proceedings of the sessions after the fact was usually a low priority for the participants and the chair, who had gleaned the benefit of the meeting already. By the time George

Adelman had managed to cajole the chair of the session into submitting a manuscript acceptable for publication, a couple of years had sometimes passed and the topic had lost its timeliness. More than one work session was in fact never reported out.

Schmitt's ultimate fantasy, of course, was a reprise of the success he had experienced with the Biophysical Study Program in 1958. He was determined to do for neuroscience what he had done for biophysics. As the midpoint of the 1960s approached, Schmitt set the summer of 1966 as the target date for a return to Boulder where another elite set of scientists would convene for the first Neurosciences Study Program.

[1]   Most of the information on Schmitt in this book is based on his autobiography: Schmitt, F. O. 1990. *The Never-Ceasing Search*. Philadelphia: American Philosophical Society. Additional details and perspectives about Schmitt are given in Irwin, L. N. 2007. *Scotophobin: Darkness at the Dawn of the Search for Memory Molecules*. Lanham, MD: Hamilton Books.

[2]   Schmitt, F. O. 1990. *op. cit.*, p. 63.

[3]   Schmitt, F. O. 1990. *op. cit.*, p. 201.

[4]   Schmitt, F. O. 1962. *Macromolecular Specificity and Biological Memory*. Cambridge: MIT Press.

[5]   Schmitt, F. O. 1990. *op. cit.*, p. 223.

8

Becoming a Scientist

In 1962, Eric Kandel began his work with Ladislav Tauc on *Aplysia* in Paris, Holger Hydén showed base changes in nucleic acids in brain cells of rats taught to balance on a wire, Gerald Edelman and Joseph Gally isolated the light chain of a purified antibody, Marshall Nirenberg and Severo Ochoa competed in a frenzied race to decipher the genetic code, and Frank Schmitt created the Neurosciences Research Program. My scientific career began that year as well, but in a place and manner more different from that of those pioneers of neuroscience and molecular biology than I could have imagined, had I known about them at the time.

Hard Days and Hot Nights at the Elbow Room

Southwest of Lubbock, the flat plateau of the caprock folds slightly into barely rolling hills. The soil is soaked throughout the spring and summer from deep-water wells that turn the land to emerald fields of cotton by early fall. As the caprock breaks down into the Permian Basin, the land becomes too dry even for irrigation. Water wells give way to oil jacks, their heads bobbing slowly up and down like a praying mantis feeding on a captured meal. The cotton fields turn to seas of dry grass and scattered mesquite, then to scrub brush and desert as the highway

veers west toward the right angle elbow of the Texas–New Mexico
border, just north of the small community of Kermit, at the eastern edge
of the Chihuahuan Basin, 153 miles southwest of Lubbock.

It was here that Don Tinkle had set up study areas on the population
dynamics and ecology of the side-blotched lizard, which occupied the
desert in greater numbers than any other vertebrate in the region of
Kermit. For many years, including the four that I was at Texas Tech,
Tinkle spent nearly every weekend and most of every summer capturing
lizards by the hundreds, marking them, recapturing them to measure
their growth, position, reproductive status, and anything else that helped
specify the animal's place in nature. The work was as hard as the land
was harsh. On an average day in July, the sun shines brightly on Kermit
for fourteen hours. It rises in the cool of the morning, prompting gentle
breezes, but soon climbs high to heat the air, bake the sand, and banish
the clouds that give relief. By three in the afternoon, the air temperature
nears 100 degrees Fahrenheit, and that of the sand 40 degrees higher.
Only at dusk, after often blazing sunsets, do the breezes return to cool
the desert down.

I was first hired by Tinkle to work in Kermit the summer following
my freshman year. I worked there the next summer as well and most
weekends of my sophomore and junior years. During the week, I worked
in the labs in Lubbock or on data in Tinkle's office. But my days were
also crowded with classes, and my evenings were filled with comparative
anatomy, organic chemistry, and the other subjects that I needed for a
chemistry major and zoology minor. The pace was relentless, and the
main thing I remember about those early college days was being tired
all the time. The education and experience came in chunks, too fast to
enjoy and too big to digest. But I kept at it because I wanted to get to
graduate school as soon as possible, to get on with my mission in life.

Through all the work with Tinkle on population ecology and all
my nitty-gritty courses in chemistry, my ultimate interest in behavior
persisted. During my second summer in the desert, Tinkle agreed to let
me work during the day on a project of my own design—a field study
of the social behavior of the lizard population that was the focus of our
research. This entailed long hours in the hot sun, sitting, crawling, or

lying on the burning sand to follow the movements and interactions of specific lizards (which on most days were monotonously minimal).

The rigor of this lifestyle did have its social compensations. Often, after a hard day of hours in the desert heat, we retired to the neighborhood tavern—the Elbow Room—for cold beers, conversation, and after rehashing the fulfillments and frustrations of the day, an endless shuffleboard tournament that Tinkle and I as partners eventually came to dominate. The Elbow Room did much to dissolve the aches of our daytime labor. Likewise dissolving were distinctions between students and professors, desert biologists and oilfield roughnecks, townspeople and scholars. To be a part of the world that was Kermit, yet a part of the world beyond it, and to be both a friend and apprentice simultaneously was a combination of experiences that I would seek forever after, and too rarely find.

On August 8, 1963, I spent my 211th hour of observation of a free-moving lizard population in the desert, surpassing the previously published record. By the end of the summer, I had amassed enough data to keep me busy analyzing it for a year. By the time I was a senior at Texas Tech, I had written my first scientific paper, which was published in June of 1965. It stands today still as one of the most comprehensive studies of daily activity and social interaction in any reptilian species.[1] It continues to be cited from time to time, bringing back memories of the long, hard, hot summer in Kermit, where, at the age of twenty, my perception of what it takes to succeed in science was formed.

Kay Lanette Thornton

In the spring semester of my freshman year, Kay Thornton and I gradually became aware of each other. Soft-spoken, thoughtful, slow to speak up—she sat in the backmost row of my English class. I, on the other hand, sat on the front row, anxious to share my views on anything as often as our instructor, Kenneth Davis, would allow. We began with suspicions of one another: she, the consummate humanist, and me, the zealous scientist, both of us filled with all the conviction that nineteen years of wisdom can engender. But our suspicions of each other dissolved over time, as we found ourselves in surprising agreement on points of literature and philosophy more often than either of us expected. I didn't think of her as beautiful in the beginning, though her dark hair and brown eyes certainly attracted me. There was a radiance about her though, especially when she was thinking intently or mischievously, that was very alluring.

In the beginning, it was her intellect that captured my attention. English was her major, and she was already familiar with a much broader literature than I. We had fundamentalist religious backgrounds in common (she, Church of Christ; me, Southern Baptist); though, like me, she had begun to explore religious notions beyond the boundaries of her upbringing. She appreciated and was interested in the natural and social sciences, as well as humanities. Her professed fascination with the spadefoot toads that had come out in hordes following a thunderstorm during one of our early dates went a long way toward winning my heart. She of course enjoyed talking to me about the world of literature that I was in the process of discovering. For my part, I confess that I was drawn also by the touch of aloofness and iconoclasm that she projected because I was feeling much the same way at the time. We had a Coke date by the end of the semester and began corresponding by mail during my first summer of fieldwork in Kermit.

We got to know each other well, we thought—first through our letters that summer, then in person as our sophomore year unfolded. We talked of ideas and shared aspirations, including my visions of a career in science and her love of literature and writing. We thought of ourselves as different from the students around us—more serious, more committed, more interested in the world beyond West Texas. We reinforced these self-perceptions, accurate or not, and fed each other's sense of destiny. With winter approaching, our friendship evolved into romance—to the extent that our fundamentalist religious mores allowed. The grind of coursework and jobs left us little time together, and the frustration of our separation grew more serious. While marriage was the last thing I needed at such an early age with such a long road ahead, the notion of a lifelong companion with whom I felt so intellectually and emotionally compatible began to gnaw at the back of my mind.

It is noteworthy that several of the men that I had shared the labor and camaraderie of fieldwork with the previous summer had combined early marriage as students with successful pursuit of demanding scientific careers, but I was not unmindful of the perils of a youthful marriage. On balance though, I began to think that the stability of constant companionship seemed preferable to the frustration of nearly constant separation from such a close, kindred spirit, who understood and supported and would help me sustain my vaunted ambitions. So in early December, I asked Kay if she would consider marrying me. She responded with cautious enthusiasm. We mulled it over in great detail on several occasions, exploring as many of the implications and potential problems as our limited experience enabled us to envision. We decided we could do it, though at first we thought we should wait at least a couple of years. By the time we became officially engaged just before Christmas, however, waiting that long seemed to serve no useful purpose; so at the end of January 1963, between finals for the fall semester and the start of classes for the spring of our sophomore year at Texas Tech, we were married.

Our first year together was occasioned by the typical bumps that occur at that stage, but nothing serious enough to shake our affection for each other or the conviction that we had done the right thing. We were proud of each other and happy to be beyond the dating game. The summer of that year, I was in Kermit watching lizards for hours on end, and the strain of that separation was felt by both of us. I couldn't wait to get back to Lubbock every two weeks or so and hated when I had to leave for the desert again. But with the fall semester, our domestic lives returned to normal. The assassination of President Kennedy, whom we both admired, was a terrible shock that depressed us for weeks. But we managed to reach the end of the year, comfortable and happy with each other.

The first serious crisis for us arose in March of 1964. Perpetually fatigued from our work and courses, we both had looked forward to the spring break for some badly needed downtime together. But at the last minute, Tinkle drafted me to go with him on an expedition to some isolated islands in the Gulf of California, and the time together that Kay and I had longed for vanished. She was brave about it, but gravely disappointed, and I was excited enough about the coming adventure that I didn't take note of her displeasure as much as I should have. When I got back, we talked it out and said we would move on. In any event, I felt confident that the coming summer would make up for the stress we had endured from being apart so often.

Summer of '64

The National Science Foundation (NSF) over the years has funded a marvelous program that enables undergraduate students to become involved in research early in their academic careers. I knew about it because Texas Tech had applied for one; but I didn't want to stay in Lubbock (much less, Kermit) that summer, so I wrote to NSF for their list of grantee institutions that year. The two programs that sounded closest to my interests were at the University of Chicago and the University of Kansas. Both responded to my inquiries, though I was particularly impressed that the chairman of the department at Kansas, Dr. Frederick E. Samson, had written a personal letter responding to my specific questions. I applied for summer fellowships at both Chicago and Kansas. Most institutions strongly favored their own students, so I was not surprised to be turned down at Chicago, but was elated when Kansas decided to take a chance on me. Kay seemed happy enough to go with me, but her best friend was expecting a baby any day, so she decided to join me later. With finals mercifully out of the way, I boarded a bus the first evening in June for the eighteen-hour ride to Lawrence and a fateful summer of work in the lab of Fred Samson.

I arrived in Lawrence on a cool, rainy afternoon with one heavy suitcase, little money, and no place to stay. I found a room for seven dollars a week, where I spent most of the next three days reading about energy metabolism and the brain—the focus of Samson's research. On the fourth day, I went to meet the author of the work himself.

Frederick E. Samson was not quite what I had expected—a little shorter in stature, decidedly less professorial looking, and a lot more talkative. He showed me through the labs, assigned me to a desk of my own, and talked with great animation about the work that he was doing. To my delight, I was well enough prepared to understand what he was talking about and to engage in a real give-and-take conversation. By the end of our first session together, I suspect he was feeling relief that the chance he had decided to take on the outside student from Texas Tech was probably not going to come back to haunt him.

My second day on the job, Samson showed me a recent paper that claimed to reverse the anesthetizing action of procaine through the application of high-energy compounds such as adenosine triphosphate (ATP). Samson had a lot of experience with ATP, but this was a highly unusual claim, so he suggested that I try to replicate the experiment. With Frank Scammon, an electrical engineering student also working in the lab, we set up a nerve conduction experiment using the vagus nerve of a turtle. Frank knew all the electrical instrumentation, and I knew enough biochemistry and physiology to handle the solutions and tissues, so we were running actual experiments by the end of the week. It was hard work, but I was so excited I could hardly contain myself. Away from the West Texas plains for less than two weeks, and I was already doing big-time research in neurophysiology! On Friday, after a week of work in Kansas, Samson reviewed what we had done and said it was good.

The days rolled on. Every morning I awoke with a sense of mission and adventure. I couldn't wait to get to the lab. And I was good at what we were doing. I couldn't have managed the electrical instrumentation on my own, but I learned from Frank quickly and well. This was so different from the painstaking zoological fieldwork, where it took at least a summer to amass some data that might or might not fit into some coherent pattern. By contrast, a single day's work in neurophysiology conceivably could result in a clear answer to a precise question. (And not incidentally, it could be done in air-conditioned comfort.) So every day I awoke to the possibility that this might be the day of the breakthrough—the big discovery that would tell us something new and fundamentally important about the function of nerve cells.

Frank seemed to enjoy our work, but he wasn't mystical about it like I was. It was a job for him, and when it was quitting time, it was time to quit. My quitting time was whenever the experiment naturally ended, and that could be well past the dinner hour. Frank upset me greatly on a couple of occasions when he stopped in the middle of an experiment simply because it was four o'clock. I worked right through lunch as often as not and would work all night willingly if that were what it took. I couldn't understand, and had no patience for, the mentality that science was open for business only certain hours of the day. Tinkle would have scoffed at that, and I figured that Samson would too. I just hadn't been around indoor scientists that much.

So I was really living my dream. But for Kay, it was a different matter. I had come to Kansas two weeks ahead of her, and the separation had left me missing her more than I had expected. After she arrived, though, I failed to convey that; and my working at the lab till all hours seemed to speak otherwise. I was living out what I had said that life with me would be like, and I thought that she would relish the opportunity to work on her writing projects with the same fervor that I had for my activities. What she really needed, though, and probably always had wanted, was more companionship instead of more work.

The summer of '64 was a turning point—in my scientific interests and career, in our young marriage, and on the national and international stage. Congress passed the Civil Rights Act, and Bob Dylan released "It Ain't Me, Babe." In mid-July, the Republican Party nominated ultraconservative Senator Barry Goldwater to run for president; and in his acceptance speech, he declared that "Extremism in the defense of liberty is no vice." Lyndon Johnson, his administration less than a year old following Kennedy's assassination, was anxious to be elected president in his own right and was not about to let Goldwater label him hesitant in resisting the communists in Laos and Vietnam. Thus in March, he had authorized covert anti-communist raids into Laos, and Pentagon planners had begun to pinpoint potential bombing targets in North Vietnam. Clandestine actions against North Vietnam, organized and aided by American agents, had been going on for years, though this was little known by the public or by most members of Congress. The night of July 30, South Vietnamese commandos attacked two offshore islands near the Red River delta of North Vietnam. Three days later, the US destroyer *Maddox*, assigned to the Gulf of Tonkin to monitor the radio traffic generated by the commando raids, came within ten miles of the Red River delta and was attacked by North Vietnamese speedboats.

President Johnson rejected the call of his advisers to retaliate, but ordered the *Maddox* and another destroyer, the *C. Turner Joy*, to respond with force to any hostilities directed against them. The night of August 3, South Vietnamese commandos staged raids again. The resulting radio traffic gave the *Maddox* and *Turner Joy* the impression that the North Vietnamese were about to strike back. With their nerves on edge, and with radar blips from a stormy sea interpreted as incoming torpedoes, the crews of the two destroyers began firing wildly into the black night. Though no confirming evidence that the ships were actually attacked ever emerged, Washington interpreted the incident as "a second deliberate attack," and this time, President Johnson ordered retaliation.

Frank and I had a radio set up in the lab where we were conducting our experiments. We had followed the escalating confrontations in the Gulf with avid interest, expecting a US response. I arrived at the lab on Wednesday morning, August 5, to learn that air attacks had been carried out against four North Vietnamese patrol boat bases and an oil depot. An estimated twenty-five patrol boats were sunk or damaged, and two of the sixty-four US planes were shot down. It seemed like a pretty heavy-handed response to me, but I admit to feeling more excited than critical. The afternoon of August 6, Congress passed the Tonkin Gulf Resolution authorizing the president to "take all necessary measures" to "prevent further aggression." Only two senators voted against it—Wayne Morse of Oregon, who labeled it a mistake of historical proportions, and Ernest Gruening of Alaska, who stated that "all Vietnam is not worth the life of a single American boy." The House of Representatives approved it unanimously.

Like everyone interested in world affairs at the time, I was aware of the fighting in Southeast Asia and knew that we were involved in a peripheral way; but also like most Americans, I was ignorant of its historical roots and sociopolitical context. I was aware that, unlike the Korean conflict, these hostilities had smoldered into being gradually and that our involvement slowly deepened by bits and pieces. But it wasn't until the air raids in the Gulf of Tonkin on August 5 that the war in Vietnam became a reality for me. And had I known what lay ahead that August morning as talk of war filled the airwaves, I surely would have been too depressed to work. As it was, there was hope expressed by some, and I guess I shared the view that perhaps the North Vietnamese would shrink in the face of the overwhelming American firepower and the incipient war would wither away.

In the end, the summer fell short of my expectations in a couple of important respects. Despite the optimism of our exciting beginning, the more data we collected as our research project progressed, the more unclear the results became. Nor did Kay and I enjoy the restoration that our time together should have provided. The unrelieved heat sapped her energy, and my absence either added to or failed to check her depression. So the summer ended in frustration for both of us.

On the other hand, there were two very positive outcomes of the summer for me personally. First, I confirmed to myself that I could do physiological research and that I really enjoyed it. Secondly, I had come to see in Samson a scientific style different from that of Tinkle, thus broadening (or tempering, perhaps) my view of what a scientist should be like. Samson appreciated long hours and hard work, but effort alone was not its own virtue. There had to be a good idea behind the effort. Imagination counted with him. Maybe imagination counted to a fault, but it was a welcome correction to Tinkle's relentless glorification of toil and sweat.

Kay and I left Kansas the second week in August to take a trip to Chicago to see friends who had moved there from Lubbock and to lift our spirits. Chicago was the biggest city I had ever seen, and I was awed to speechlessness by the taxi ride along Lakeshore Drive. The palisade of skyscrapers, the density and bustle of traffic, and the noise and odor and energy of an urban sprawl more vast than any I had ever imagined, overwhelmed my senses. In our few days there, I was amazed by other novel experiences: riding on a subway, seeing a major league baseball game, encountering locked doors to apartment buildings during broad daylight, and feeling the crispness of cold air from Canada that dropped the daytime temperature below 80 degrees before the first of September. If I ever had a doubt about needing to leave Texas, the trip to Chicago left me knowing that a larger world was out there, waiting for my arrival. But first, we had another year to go at Texas Tech

Winter of Discontent

I was pleased to get back to Lubbock and to Tinkle to relate to him the adventures and insights of the summer. Having seen the broader world, I no longer felt confined by the cotton fields and oil derricks of West Texas or the parochial insularity of Lubbock. I knew the day was coming when I would be leaving; and having tasted research of a different sort, I felt that the elements of an exciting career in science were falling into place. Graduate school was on the horizon, and the time to apply had come.

From a list of over twenty possibilities, I paired my choices down to four by October: Western Reserve, Yale, the University of Pennsylvania, and the University of Kansas. Western Reserve had impressed me because of thoughtful letters I had received from David Cohen and Robert Josephson in response to my inquiries about graduate research on learning and memory. Yale appealed to me, in part because of its exotic location in faraway New England, but mainly because of encouraging correspondence from Robert Galambos and Daniel Kline. In late October, I had written to Galambos that in the two years since he had last written me in response to some neurophysiological questions I had posed,

> My interests have matured and gained perspective. I feel that I would like to concentrate on a physiological

approach to information storage and retrieval in the nervous system. It seems that for a breakthrough in this area, an integration of existing biochemical, EEG, ablation, etc. . . . or perhaps an entirely novel approach, is needed.

Galambos thanked me for my "thoughtful and informative" letter and added, "I think you could do much worse than to come to Yale for the program you outline . . . Whether or not you decide to come to Yale, I sincerely hope that you will be able to do work you want to do." Kline, Director of Graduate Studies, said my background seemed to be adequate and, if my graduate record exam (GRE) was good, my course grades should not "be a severe handicap to your being accepted by us or any other good graduate school."

My top choice, however, was the University of Pennsylvania and its Institute of Neurological Sciences. Among the faculty were Louis Flexner, the pioneer in drug inhibition of memory; Eliot Stellar, a noted physiological psychologist; and Vincent Dethier, a well-known invertebrate neurobiologist.

The University of Kansas was, of course, a safety school. I had enjoyed my experience there tremendously and had no doubt that the graduate education there would be fine. But I really wanted to go somewhere new, to work with scientists whose research was more behaviorally oriented and closer to the exciting developments that I knew were probably just around the corner.

The graduate school application process was time-consuming and tedious, but that was only one of the obstacles I had to get through in my final year at Texas Tech. The academic challenge was formidable, with coursework in animal physiology, physical chemistry, chemical literature, and a couple of liberal arts requirements still to complete. The pressure to make good grades, to help me get into graduate school, was greater than ever. And Kay and I both still had to work to keep from sinking further into debt. She, of course, had her own set of pressures. She, too, was applying to graduate schools (in English), and her academic schedule was heavy. Unlike me, she was an outstanding student—ultimately the winner of a Woodrow Wilson Fellowship (in part because she impressed the panel as an English major with her surprising knowledge of molecular biology)—but the demands of coursework on top of household chores and work obligations seemed incessant.

With both of us immersed in these struggles to endure our senior year, we had little time for quiet reflection and communication. The situation was aggravated, ironically and with the best of intentions, by fervent relationships with our closest friends. Upon our return to Lubbock from Kansas, Kay and I had re-immersed ourselves in the close-knit group of semi-misfits like ourselves with whom we had become associated since our earliest days at Texas Tech. In no small part, it was a reaction to the solitude of Lawrence, where we (Kay, I think, especially) had felt cut off from the people who gave us sustenance. But they also demanded a lot, and that particular semester, they overwhelmed us with trials and tribulations that seemed to know no bounds. One or another of them was in our apartment almost constantly. At all hours of the day or night, we found ourselves in endless conversations about one personal crisis after another. We all cared for one another deeply, and tempted though we were, throwing our best friends out into the night was not something that Kay or I could do.

The friendships turned from an apparent blessing to an obvious burden. It was hard enough for me to keep up with physical chemistry and work for Tinkle, much less find time to deal with the tedious problems of friends who at times seemed almost to revel in their maladjustments. An insidious cycle set in for me, in which I never quite finished my homework or resolved anyone's problems or got enough rest or completed any project with satisfaction. I began to resent the fact that I wasn't brilliant enough or energetic enough to handle it all and began to regress into the angry rebel of earlier years. Had I been able to direct my frustration more productively toward dealing with the truly critical obligations in my life, I may have weathered the tidal wave that was about to swamp me. I battled to keep my sensitivity. Maybe I helped my friends in minor, temporary ways. But sensitivity to Kay, the one who mattered most, slipped away.

Like the ominous vibrations before an earthquake, I began to sense that something was terribly wrong between Kay and me near the end of October, but I just couldn't focus on it enough in time. One morning in early November, in the course of a conversation that started benignly enough, Kay finally brought herself to confess that she felt that we had grown too far apart to stay together. In an instant, I realized what we had allowed to happen. Filled with remorse, I scrambled to hold the walls of our marriage together, and she said she would try too. For a couple of weeks, we teetered on the brink of reconciliation, but the neglect had been a little too deep for a little too long. For Kay, there was an emotional component that I had failed to bring to my vision of our bright future together. That I felt it, didn't matter; that I failed through my deeds to express it, did.

A couple more experienced with the frustrations of young adulthood or more patient with the ebb and flow of human feelings could have found a way to survive such a crisis. But we were barely twenty-two, and the prospect of a lifetime together but unhappy was more frightening than the uncertainty of going our separate ways in time to reconstruct our respective dreams.

How I managed to stagger to the end of the year, get through the holidays with our two families, and survive my final exams in January is still a wonder. Incredibly enough, I made grades of A in chemical literature and animal physiology, and a C in physical chemistry (for which I was most grateful). With an A and a B in my liberal arts courses, I pulled my grade point average within striking distance of a level high enough to merit my admission to graduate school, or so I thought.

I began the spring semester of my senior year with a heavy heart, as it appeared increasingly unlikely that Kay and I were going to find our way back to each other. A slight counterpoint to the grave disappointment in my personal life was the anticipation of acceptance to graduate school and the certain knowledge that, whatever the fate of my marriage, by September, I would be out of Texas and in a graduate school somewhere.

By the third week in January, I had my first acceptance in hand from the University of Kansas. Though my home department could not commit to a teaching fellowship until March, I wasn't too worried. On March 22, I received a letter of rejection from Yale. Notwithstanding Galambos's encouragement and Kline's speculation to the contrary, my grades apparently weren't good enough for Yale. The rejection would have been more discouraging had Kay not been rejected as well. Since she was a Woodrow Wilson Fellow, it was easy for me to conclude that I simply had been overambitious in hoping for Yale. Kay did get accepted by the University of Pennsylvania, so I continued to hold out hopes for my first choice.

It was not to be. On April 4, I returned to Lubbock from spring vacation to find a letter of rejection from John Brobeck, Chairman of Physiology at the University of Pennsylvania. As if to make certain that I got the message, there was also a form letter from the vice dean of the Graduate School regretting to inform me that I could not be admitted. It was addressed "Dear Applicant." Finally, about a week later, Eliot Stellar wrote to inform me that since I had not been accepted by my home department, there could be no place for me in the Institute of Neurological Sciences. This was a crushing blow. I knew that Flexner, in particular, was doing just the type of work I wanted to do. I never heard a word from Flexner.

By the time my letter of rejection from Western Reserve had arrived, I had come to expect it, so was just more sad than shocked. By now, I was beginning to worry after all about financial aid even from Kansas. The day after my rejection from Western Reserve, my fears were realized when I received a letter informing me that no assistantships were available for any first-year student in my home department at Kansas. The destruction to my ego was near complete. From the high aspirations and hard work of the fall and winter application process, I had managed to get into only one graduate school—my fourth choice of four—and with no financial aid even there.

By April of 1965, then, my life had taken on a melodramatic character of epic proportions. But there was more. Kay and I had always wanted to have children, so in the autumn, before our estrangement, we had conceived a child with earnest aspirations and great intentions. By the time we knew that Kay was pregnant, we had decided, with no obvious justification to an outside observer, that our attempt at reconciliation was past the point of no return. On June 2, 1965, our son, Michael Sean, was born. It was a bright beautiful day in West Texas, and I remember walking across the campus in the brilliant early summer sunlight, an hour late to my registration for summer school because of the morning's drama, with the most powerful set of contradictory emotions I had ever felt: excitement, pride, anxiety, and remorse—all churning together in chaos. The enormity of what we had done, at both the positive and negative ends of the spectrum, was more than I could comprehend.

Gravely disappointed at the outcome of my aspirations for graduate school, I was filled with anxiety about my professional future. I was heartsick over the breakup of my marriage and angry at the stupidity that had let it happen. But more than anything that summer, and for many years to come, I worried most about bringing a child into the world, the product of earnest and caring but confused and self-absorbed parents. Had I been able to foresee the power and effect of his mother's love or the talent and character that he would summon beyond the bounds of either heredity or environment, I would have worried much less. But in those days, I hardly knew what was happening in the present, much less what the future would hold.

Lubbock in My Rearview Mirror

My roller-coaster ride of great promise, deep disappointment, love and marriage, loss and regret, new beginnings, and—somehow—ambition still intact was thankfully coming to an end in Texas. Like Mac Davis, I couldn't wait to see Lubbock in my rearview mirror.[2] With two more courses to finish in order to graduate in August, I settled in to survive one last summer on the West Texas plains.

Dorothy Haecker, my lab partner in high school physics, and I had become good friends before we had headed in different directions to college, but had kept in touch off and on throughout our undergraduate years. So it was not unnatural for me to share my sad news with her during the spring semester. That led to stepped-up correspondence that revealed that she too was going through some emotional turbulence of her own. This recharged our friendship and provided us both with some badly needed solace as the summer progressed.

Another friend in need was Penny May, my classmate from freshman biology, who had remained my best new friend besides Kay throughout our time at Texas Tech. Challenged by a situation beyond her control, she had needed a confidant in the worst way, and I had tried to be that; but enveloped in my own personal drama, I had failed to provide enough of the support that she most needed at the height of her calamity. She was dismissed from Texas Tech, truly through no fault of her own, for the spring and summer, returning to finish her degree after I had left the following year.

I signed up for early morning sections of my two remaining courses—English literature and Texas history—so that I could devote my afternoons and evenings to working for Tinkle and reading the literature on learning and memory. I was already well acquainted with the work of Hydén, Flexner, and Rosenzweig and his colleagues. The stream of publications from Agranoff and Barondes was in its early stages, so I absorbed their output, read Karl Lashley's classic *Brain Mechanisms and Intelligence*, and spent a lot of time getting caught up on the experiments of James McGaugh. Besides his experiments on memory disruption by electrocortical shock, he had also used stimulant drugs like picrotoxin[3] and strychnine[4] to *facilitate* memory consolidation. In every case, the effect of the stimulants was time dependent, with decreasing effect when administered at longer intervals after training. This helped reinforce the growing evidence for multiple stages in memory deposition. In the Soviet Union, Leonid Pevzner was showing, as Hydén had done before him, that a variety of methods for stimulating discrete regions of the brain could alter the rate of turnover in RNA.[5] Taken together, all these results pointed to a role for synthesis of new RNA and/or protein that a new behavioral experience set in motion, but that wasn't completed until a molecular storage form of long term memory had been laid down.

With this as backdrop, the scientific world was primed for a dramatic leap in the connection between molecules and memory. Thus, with many others, my jaw dropped when I opened my August 6, 1965, issue of *Science* and read a claim by a group of previously unknown investigators that they had induced untrained rats to approach a food cup more readily at the sound of a click if they had been injected with RNA from the brains of rats previously conditioned to do so.[6] A less dramatic but ultimately more consequential report of which I was unaware at the time was a short paper from a group at the Baylor College of Medicine that actinomycin D was able to inhibit the well-known development of tolerance to repeated administration of morphine.[7]

I managed to learn enough English literature and Texas history to pass my last two courses at Texas Tech and graduate with a bachelor of arts in chemistry on August 21, 1965. On the first day of September, I crossed the Red River, leaving Texas for the adventure that awaited me in Kansas and beyond as the 1960s reached their midpoint.

[1]   Irwin, L. N. 1965. Diel activity and social interaction in the lizard *Uta stansburiana stejnegerii. Copeia* 1965: 99–101.

[2]   © Mac Davis, 1980, "(Lubbock) Texas in My Rear View Mirror," Casablanca Records. Davis was a Lubbock native.

[3]   Breen, R. A. and McGaugh J. 1961. Facilitation of maze learning with posttrial injections of picrotoxin. *J Comp Physiol Psychol* 54: 498–501.

[4]   Petrinovich, L F., D. Bradford, and J.L. McGaugh. 1965. Drug facilitation of memory in rats. *Psychon Sci* 2: 191–2.

[5]   Pevzner, L. Z. 1965. Topochemical aspects of nucleic acid and protein metabolism within the neuron-neuroglia unit of the superior cervical ganglion. *J Neurochem* 12: 993–1002; Pevzner, L. Z. 1966. Nucleic acid changes during behavioral events. In *Macromolecules and Behavior*, edited by J. Gaito. New York: Appleton, Century, Crofts.

[6]   Babich, F. R., A. L. Jacobson, S. Bubash, and A. Jacobson. 1965. Transfer of a response to naive rats by injection of ribonucleic acid extracted from trained rats. *Science* 149: 656–7.

[7]   Cohen, M., A. S. Keats, W. Krivoy, and G. Ungar. 1965. Effect of actinomycin D on morphine tolerance. *Proc Soc Exp Biol Med* 119: 381–4.

LATER SIXTIES
(LAWRENCE)

9

Transfer of Memory

Studies on the *encoding* of memory and the *disruption* of memory were well underway by 1965, but the strategy of *transferring* memory from animals that had learned a task to naive (untrained) recipients by injecting extracts of the brains of trained donors appeared to come out of nowhere that summer. In fact, that wasn't quite the case. In the years that followed, two names would become most closely associated with the transfer phenomenon. One was *Jim McConnell*, a psychologist at the University of Michigan who had begun with the modest objective of studying learning, as Kandel had, in an invertebrate with a simple nervous system. The other was *Georges Ungar*, a neuropharmacologist at the Baylor College of Medicine in Houston, who saw an analogy between an animal's development of tolerance to a drug and its ability to store memory.

James V. McConnell

James Vernon McConnell was born in 1925 in Okmulgee, Oklahoma, but raised in Shreveport, Louisiana. He enrolled at the age of sixteen at the Louisiana State University (LSU) intending to study chemical engineering, but soon discovered psychology to be much more interesting. He served in the navy briefly at the end of WW2, then completed his degree at LSU in 1947. After a couple of jobs as a radio announcer, he landed a position with WLW-TV in Cincinnati. Commercial television was barely more than experimental in 1950. His experience in college dramatics enabled him to write commercials, skits, and all the material he used as an announcer, including the first daytime soap opera for television, which management cancelled on grounds that televised soap operas would never be able to compete with those on radio.

The stress of work at the television station gave him an ulcer that forced him to return to the family business in Shreveport, but then the death of his father threw him into a deep depression till a friend provoked him to escape his funk by applying to graduate school. He was accepted by the University of Texas at Austin, intending to study clinical psychology. This was thwarted, however, by Wayne Holzman— director of the clinical program and creator of the famous ink block test for personality evaluation—who complained about the "bad attitude" that McConnell displayed by constantly asking his professors for data to back up their assertions. McConnell decided that experimental animal psychology, where data didn't present an attitude problem, might be a better fit.

He first sought a position in the lab of Jeff Bitterman (the same Bitterman who complained years later to Agranoff at Bryn Mawr that goldfish were too difficult to train because they never unlearned anything). Bitterman, at the time, was studying earthworm behavior, but McConnell wanted to see if learning could occur in an animal with an even simpler nervous system, so with the help of Robert Thompson, a fellow graduate student, decided to try classical conditioning in planaria. This aquatic flatworm, just a few millimeters long with slightly enlarged ganglia at its head end that nominally qualify as a brain, typically lives on the underside of rocks in freshwater streams. McConnell and Thompson set up a trough of water in McConnell's kitchen with electrodes attached so they could deliver a mild shock through the water a few seconds after turning on an overhead light. Following a number of trials, worms stimulated by the light alone would show contractile reactions that previously had been induced by shock, even if no shock were delivered. McConnell and Thompson felt they had demonstrated classical conditioning in the simplest animal ever tested for its capacity to learn, and the *Journal of Comparative and Physiological Psychology* agreed.[1]

Bitterman was dubious of the qualitative nature of the criteria used to claim conditioning in flatworms, so McConnell abandoned his worm research until a few years later after completion of his PhD degree. Upon being hired in 1956 as an assistant professor of psychology at the University of Michigan, it was emphasized to him that he would have to publish or perish. Since most of his colleagues were concentrating on the main subjects of academic psychology departments—white rats and college sophomores—he decided to return to the question of learning in animals with very simple nervous systems, where he would have the field of invertebrate learning at Michigan to himself.

His thoughts returned to a conversation with Thompson a few years earlier, when his colleague had pointed to a magazine ad for Toni Home Permanent, showing a picture of twin girls—one with her hair done at the beauty shop while the other had hers done equally well with the Toni product at home. You know, Thompson pointed out, when you cut a planarian crosswise in half, the head end regenerates a tail and the tail end regenerates a head, so you end up with two new and indistinguishable worms. If that were done to a planarian conditioned to react to a light stimulus, would the tail section that regenerated a head retain the memory for the conditioned response as well as the head section that regenerated a tail? It was an experiment they never got to perform in Austin, but now McConnell thought he would give it a try in Ann Arbor.

He began to recruit students to his cause. With Dan Kimble, a first-year graduate student, and Allan Jacobson, an undergraduate honors student, the long-delayed experiment was carried out. The worm regenerated from the tail section was found to retain the conditioned response as well as the one regenerated from the head end.[2] It appeared that learning was a distributed property of the whole animal. This suggested to McConnell that behavioral information had been encoded into some distributed chemical form; and this, in turn, inspired a more audacious experiment.

Upon learning that planaria could be cannibalistic, McConnell decided to feed chopped up pieces of conditioned planaria to naive worms to see if the cannibals would show evidence of conditioning without training. Apparently they did, though the differences between experimentals and the single control group were marginal.[3] Later experiments[4] appeared to be better controlled, but by then, a skeptical scientific community was pushing back, claiming the results were more likely due to sensitization, nutritional effects, or some other factors.[5]

In the face of these criticisms, to his credit, McConnell was open and anxious to have others try his experiments. Melvin Calvin's lab at Berkeley was very interested in the mechanisms of learning, so Ed Bennett was assigned to try this more dramatic approach than the methodical neurochemical and neuroanatomical studies he had been working on with Rosenzweig, Krech, and Diamond. To assist him, McConnell dispatched two of his students, Allan and Reeva Jacobson (unrelated), to Berkeley to consult with and oversee the proper procedures with Bennett. The collaboration ultimately failed, however. The Jacobsons felt they were training the planaria successfully, but Bennett could not be convinced. Whether he was too skeptical to be objective, or they were too qualitative in their evaluation of conditioning, the bottom line was a lack of agreement on how the experiments turned out. Whether this failure was a reason, or Allan Jacobson was already thinking of a bigger leap anyway, he was on the path to make an even more startling claim.

By 1965, Jacobson had completed his graduate work with McConnell and obtained a position in the Psychology Department at the University of California–Los Angeles (UCLA). There he carried out the transfer strategy to its logical extreme. He trained rats to approach a food cup at the sound of a click, then extracted RNA from their brains and injected it into the abdominal cavity of naive rats and counted the number of times the recipients spontaneously approached the food cup at the sound of a click. Jacobson and his team acknowledged the obvious difficulty of determining how specific the effect was (as opposed to sensitization or other non-specific behavioral effects), or whether substances other than RNA in the injected extracts might affect the recipients' behavior, or even beginning to explain how RNA – assuming it got into the brain – could affect brain function.[6] In another report in *Science* about three months later, Jacobson's team reported that recipients tended to respond differently to the sound of a click or a blinking light, depending on which of those stimuli had been used in training the donor rats.[7] The effect, therefore, appeared to be behaviorally specific.

The trail from McConnell's kitchen in Austin, where a decade earlier he had first tried to study learning in the simplest animal capable of possessing a Hebbian (modifiable) synapse after Wayne Holzman had complained that he was too obsessed with data in the classroom, had now culminated through his student in one of the potentially greatest discoveries in neuroscience—that memory in a neural system with the complexity of a mammalian brain could be encoded into molecules of RNA. His starting point conceptually had not been much different from Kandel's. But this line of research would have a very different ending.[8]

Georges Ungar

Born in 1906 in Budapest and raised in an intellectually stimulating household in France by his French mother and Hungarian father, Georges Ungar was initially interested in the classics. He won first prize in a nationwide Greek scholarship competition at the age of nineteen and finished second in history to a person who later became a noted historian. He was attracted first to art history, then to architecture, and finally to medicine, which his zoology professor declared was just as well, since his (Ungar's) drawings were so bad that he "would never amount to anything in science."

He didn't particularly like medical school, and his bedside manner left a lot to be desired, but he did find the challenge of diagnosing diseases something he enjoyed and was good at. The problem-solving aspects of diagnosis intrigued him and led him to question his zoology professor's judgment about his lack of aptitude for science. He sought a professor with whom he could try his hand at research and was led by contacts through friends to Jules Tinel.

France gave the world two of the nineteenth century's greatest medical scientists. One was Louis Pasteur, who proved the microbial basis of infectious disease and disproved the spontaneous generation of life. The other was Claude Bernard, whose methodical and empirical study of organ function in living animals—the essence of physiology— made physiology the queen of the medical sciences in the nineteenth century, giving it the aura and stature that microbiology and molecular biology would achieve in the twentieth. The success of physiology depended largely on its objective, empirical, and rigidly mechanistic approach to bodily functions, which at times could become excessive. A neurophysiologist and one of Bernard's students, Francois Frank, welcomed a visitor one day to his lab, where a disemboweled animal lay with tubes inserted into several veins, its heart attached to a cardiograph, and one of its kidneys enclosed in another instrument. "Now we are ready to study emotion!" proclaimed the proud Francois Frank.[9]

Tinel was a student of Francois Frank, so when Ungar went to work with Tinel, he became a direct academic descendent of Claude Bernard.[10] Tinel was working on the physiological effects of histamine. To control their experiments properly, Ungar and Tinel needed to specifically block histamine. Ungar's task was to test the ability of different chemical compounds to block histamine's effect on a strip of smooth muscle from the intestine of a guinea pig. This tissue was known to be very sensitive to the contractile effects of histamine. By measuring the amount of muscle contraction when histamine was applied to it, a quantitative measure of histamine's potency (hence amount) could be deduced. This was the principle of the bioassay, which had become a standard technique for monitoring the isolation and identification of unknown compounds or known compounds with unknown potency. The degree to which muscle contraction was inhibited by the compounds that Ungar applied in the presence of histamine was a measure of their potency as antihistamines. Daniel Bovet, a pharmaceutical chemist at the Pasteur Institute, gave Ungar a number of compounds to test, and one of them turned out to have antihistaminergic properties. This led Bovet to the discovery of generic antihistamine drugs, for which he would eventually win the Nobel Prize in 1957. In this way, Ungar contributed in the discovery of the antihistamines, which Bovet generously acknowledged.

Ungar's facility with the bioassay technique gained him recognition that began to spread internationally and would be a key to much of the rest of his career. A young pharmacist, Alberte Levillain, came to his lab in March 1937, to learn the bioassay technique. Ungar fell for her quickly and hard and persuaded her to marry him within seven months of their first meeting. For their honeymoon, they traveled to the marine biological laboratory at Arcachon, where years later, Ladislav Tauc and Eric Kandel would carry out their seminal experiments on cellular mechanisms of sensitization and habituation in *Aplysia*.

Soon after the Ungars' return to Paris to resume their research, the war intervened. Georges joined the French army, but he and Alberte had to flee south as the Nazis overran the country. They escaped from France illegally, through Algeria and Morocco to Gibraltar, and then by a harrowing voyage through German-infested waters, sailed to England to join the Free French Forces of Charles DeGaulle. Well-known in England for his research already, he was also designated a member of His Majesty's Allied Forces. The Ungars lived out the war in Oxford, where he conducted research on traumatic shock and inflammation. They intended to return to Paris after the allied victory, but Georges stayed for a year as a science adviser at the French Embassy in London. In that capacity, he was able to arrange a visit to the United States, where major advances were being made on the physiology of stress responses.

By 1948, he was well enough known in America as well to garner speaking invitations at several of the country's elite universities. While he had no intention of staying in the United States in the beginning, he received several job offers during his tour and sensed the opportunity for science that the USA had to offer. So he accepted what seemed like the best of the offers, at the Northwestern Rheumatic Institute in downtown Chicago. There he picked up where his research in Oxford on the body's response to traumatic injuries had left off. He became particularly impressed with how often irritated or injured cells caused proteins to break down, leading to an interest in biochemical responses to excitation that he would maintain for the remainder of his career.

The Ungars, who by then had an infant daughter, were dismayed to discover how "Black people, Jews, and people with small children" were not welcome in Chicago,[11] so they had to find housing in the suburban setting of Park Forest from which Georges was forced to carpool into work. They hated the cultural sterility of the suburbs, and he couldn't stand the uniquely American practice of carpooling. On top of that, the intellectual stimulation at his workplace was lacking, so they decided to return to Paris. While waiting in New York for paperwork to clear in France, Ungar was invited to help set up a lab at the US Vitamin Corporation, which he did so well that he was offered the position of directing it. He accepted and settled with his family into an apartment on the upper west side of Manhattan, which he and Alberte found much more compatible with their urbane tastes.

Ungar ended up working at the US Vitamin Corporation for eight years, where he was able to conduct research and publish in relative freedom. He earned his salary many times over when he discovered, with Seymour Shapiro, the antidiabetic drug phenformin, which, for years, was the major drug prescribed for diabetics who couldn't take or obtain insulin. During this period, he maintained his interest in proteolysis—the breakdown of protein—in response to cell damage or intense stimulation. He was aware of research in Western Europe and Russia prior to the Second World War that suggested the ability of proteins to change their shape, and he began to wonder if such changes preceded the breakdown of protein. He set out to show that prolonged neural excitation would cause conformational (shape) changes in proteins in stimulated nerve cells. Some questioned whether the techniques available at the time were sensitive enough to detect minute changes in shape at the molecular level; but Ungar insisted they did, generating a reputation for claims that, in the eyes of some, strained credulity for the rest of his career.[12]

Though happy in New York and at US Vitamin, by 1962, he was missing the academic atmosphere he had so enjoyed earlier in his research career; so he began to consider a return to institutions outside of industry. He passed up an opportunity to become program director of the National Science Foundation in favor of director of a planned Institute of Comparative Biology at the San Diego Zoo. It soon became apparent that the zoo's administration had a very different concept of what the institute should be ("Just another display," according to Ungar). In addition, he and Alberte found the religious intolerance and political conservatism of Southern California to be decidedly less appealing than the benign climate, which in part had drawn them there. So within a year, his resignation was asked for and gladly provided.

His one alternative available on short notice was a position in pharmacology at the Baylor College of Medicine in Houston. Having been warned that Texas possessed many of the less desirable characteristics of the United States, Georges and Alberte were pleasantly surprised to find a welcome and friendliness in Houston that contrasted strongly with their experiences in Chicago and San Diego. They found a comfortable apartment just a short distance from the sprawling medical center and settled in for what turned out to be nearly the entirety of their remaining careers.

At first, Ungar continued his research in Houston on conformational changes in protein. While this concept was gaining ground in conventional scientific thought, it was still too controversial for comfort so he decided to return to more conventional pharmacology. He resumed his interest in pain and analgesia (blockage of pain), which had started with Tinel all the way back in the '30s. One of the problems with analgesics like morphine (an opiate drug) is that animals develop a tolerance to it; if the drugs are administered repeatedly, higher and higher doses are required to obtain the same degree of pain suppression. A few years earlier, it had been shown that a particular peptide called substance P could reduce the analgesic effect of morphine, so Ungar decided to see if blocking peptide synthesis could increase its effect or at least diminish the development of tolerance to the drug. He used actinomycin D to block RNA synthesis and thereby prevent the synthesis of new peptides or proteins. The experiment worked; after four days, 64 percent of mice injected with a control saline solution had developed morphine tolerance, but only 31 percent had done so if injected daily with actinomycin D.[13]

Ungar then took the next logical step. If the production of a new peptide diminishes morphine tolerance, then animals already tolerant should possess more of the tolerance-resistant peptide in their brains, and extracts of their brains should *transfer* that resistance to tolerance to recipients not previously exposed to morphine. From morphine-tolerant rats left over from the previous experiment, he prepared brain extracts and injected them into naive mice, with results confirming his hypothesis—the naive mice showed a greater resistance to morphine tolerance than mice injected with brain extracts from nonmorphine-tolerant donors.[14] While dramatic, this result was obtained by the conventional bioassay method that had made Ungar an internationally recognized researcher.

Ungar's ultimate objective with these experiments had been to determine if the brain produces an endogenous antagonist to morphine in areas of the brain that sense pain. Had he continued to pursue that question, he may have discovered endogenous opiate receptors five years before several other labs did in the early 1970s. But instead, he made the fateful decision to redirect his attention toward behavior. The development of drug tolerance is a form of habituation, which, in a behavioral context, means the loss of a behavioral response to repetitions of a stimulus that has no consequences. Ungar decided to test whether this simple form of learning could be transferred in the same way that habituation to morphine had been transferred.

The stimulus Ungar selected was a sharp, loud sound of a hammer striking a metal plate. Rats will show a startle reflex the first time they are exposed to such a sound, but gradually stop reacting as the same sound keeps being repeated. When brain extracts from rats that had been fully habituated to the sound were injected into naive mice, the recipients habituated to the sound much more rapidly than mice injected with brain extracts from nonhabituated donors. And the way the extracts of the donor brains had been prepared suggested that the principal components of the extracts were likely small proteins or peptides.[15]

[1] Thompson, R. and J. V. McConnell. 1955. Classical conditioning in the planarian, *Dugesia dorotocephala*. *J Comp Physiol Psychol* 48: 65–8. McConnell himself later pointed to documentation of the learning ability in planaria in a Dutch journal thirty-five years earlier (P. Van Oye. 1920. Over het geheugen bij de platwormen en andere biologische waarnemingen bij deze dieren [About the memory of the flatworms and other biological observations in these animals]. *Natuurwetenschappelijk Tijdschritt* 2: 1–9.

[2] McConnell, J. V., A. L. Jacobson, and D. P. Kimble. 1959. The effects of regeneration upon retention of a conditioned response in the planarian. *J Comp Physiol Psychol* 52: 1–5.

[3] McConnell, J. V. 1962. Memory transfer through cannibalism in planarians. *J Neuropsychiat* 3 (Suppl. 1): S42–S48.

[4] Jacobson, A. L., C. Fried, and S. D. Horowitz. 1966. Planarians and memory. *Nature* 209: 599–601.

[5] Hartry, A. L., P. Keith-Lee, and W. D. Morton. 1964. Planaria: Memory transfer through cannibalism reexamined. *Science* 146: 274–5.

[6] Babich, F. R., A. L. Jacobson, S. Bubash, and A. Jacobson. 1965. Transfer of a response to naive rats by injection of ribonucleic acid extracted from trained rats. *Science* 149: 656-7. This paper happened to appear immediately following a paper by Hydén and Lange, extending their argument for a reciprocal metabolic relationship between neurons and glia, as discussed in an earlier chapter.

[7] Jacobson, A. L., F. R. Babich, S. Bubash, and A. Jacobson. 1965. Differential-approach tendencies produced by injection of RNA from trained rats. *Science* 150: 636–7.

[8] McConnell's life and career, including his role in planarian learning and early attempts to transfer learning, is treated in greater detail in Irwin, L. N. 2007. *Scotophobin: Darkness at the Dawn of the Search for Memory Molecules*. Lanham, MD: Hamilton Books.

[9] This anecdote was told by Ungar, whose life and career are given in more detail in Irwin, L. N. 2007. *Scotophobin, op. cit.*

[10] Since I would later work with Ungar, I, too, can trace my academic lineage straight back to Bernard. I do not, however, wish to associate my attitude with that of Francois Frank.

[11] Irwin, L. N. 2007. *Scotophobin, op. cit.*, p. 80.

[12] In time, conformational changes in membrane proteins were shown indeed to be the way that nerve cells regulate the flow of ions into and out of the cell, accounting for excitation.

[13] Cohen, M., A. S. Keats, W. Krivoy, and G. Ungar. 1965. Effect of actinomycin D on morphine tolerance. *Proc Soc Exp Biol Med* 119: 381–4.

[14]   Ungar, G. and M. Cohen. 1966. Induction of morphine tolerance by material extracted from brain of tolerant animals. *Int J Neuropharmacol* 5: 183–192.

[15]   Ungar, G. and C. OcegueraNavarro. 1965. Transfer of habituation by material extracted from brain. *Nature* 207: 301–2.

10

Graduate School at Last

From a chilly, rainy morning in April 1963, when I first crossed into Kansas in fear and apprehension at the thought of delivering my first scientific paper at the age of twenty, I have had an affection for the state that defies my usual criteria for desirable places. It doesn't have the majesty of the Colorado Rockies or the rugged beauty of Northern New England. There is no seacoast in Kansas to please the eye and no desert to test the spirit. There are the foothills of the Ozarks in the east, which have an understated attractiveness, and the vast plains in the west, which appeal to certain sensibilities. But with the possible exception of the gentle, rolling beauty of the Flint Hills just beyond Topeka, there is little in Kansas to inspire scenic wonder. The sun shines there a lot, and after years of winters on the shores of the Great Lakes and the North Atlantic, I came to appreciate that benefit; but the climate is neither mild in the winter nor cool in the summer. So it is not the landscape and not the climate that endear the state of Kansas to my mind. It is very simply that it was there that I was given a chance to make something of myself when I was deemed too risky in more prestigious places, and it was there that the patterns of the rest of my personal and professional life were laid down.

New Friends and Exciting Prospects

I settled into a cheap apartment at the base of Mount Oread, below the campus in the middle of Lawrence, and spent my time prior to the start of classes mainly reading more of the literature on learning and memory. It was during the second week in September that I came across the publication by Ungar on transfer of sound habituation in rodents.[1] This was the second amazing experiment on the possibility of molecular coding of memory that I had come across within a month. I wrote to Ungar for a reprint, which he sent immediately, signed "With compliments, G. Ungar." This token of cordiality encouraged me to start thinking about the possibility of making personal contact with such a scientist who appeared to have made a startling breakthrough on the molecular mechanisms of memory.

When I checked in with my home department, I was apprised of both good and bad news. The good news was that funds had been found to offer me a quarter time teaching assistantship, so I would get a little financial support after all. The bad news was that Fred Samson, the professor with whom I assumed I would do my graduate research, had decided to take a year-long sabbatical at an institution in Boston that I'd never heard of—the Neurosciences Research Program.[2] So consolidation of the student-mentor bond that I had so looked forward to was not going to happen for at least a year. But with coursework taking up most of my first year in graduate school, I decided not to worry about it at the time.

During those first few days in Lawrence, I also looked forward to Dorothy's arrival. As we had begun to share the outlines of the troubled personal lives we had had with others, our own personal lives had become more entwined. We didn't know where it would lead but decided that we would be better off starting graduate school together than apart, so she, too, had applied to the University of Kansas and, as a Woodrow Wilson Fellow, had no trouble being accepted by her chosen department of philosophy. She arrived in Lawrence a week after I did, and we provided valued friendship and consolation to each other during those first few weeks when neither of us knew anyone else there.

In time, I did come to form other friendships that turned out to last a lifetime. My office mate was Art Friesen, like me a first-year graduate student. Art was from a Mennonite family in British Columbia. He had gone to a small church college in western Kansas and wanted with a passion to go to medical school, but had not been admitted. Graduate school was a holding action for him. He was an excellent student and a whiz at intermediary metabolism. We became personal friends, studying together often and sharing the fortunes and frustrations of our respective ambitions. We both also indulged a pathological attraction for *Batman,* which we watched with regularity every afternoon at the student union, along with millions of preschoolers throughout the western world.

There was a poignancy to the friendship between Art and me, nurtured by the deepening conflict in Vietnam. As a Canadian citizen, Art was not subject to the draft, but in the pacifistic Mennonite tradition, he was ardently opposed to the war. Like so many Canadians who look upon the mindset of the United States with wonder and a touch of petulance, he could not fathom why we were persisting in such a misguided venture. I, of course, *was* subject to the draft, but had a student deferment. By the fall of 1965, I had growing doubts about the legitimacy of the conflict and reassured myself with the thought that what I hoped to achieve as a scientist would be much more valuable than anything I could do as a soldier. Nevertheless, for both of us, it was hard to live every day with the knowledge that men of our age were dying while we not only were safe but seemed to be making headway toward our dreams. For me, it was a discomfort that bordered on guilt, and it never went away.

Another lasting friendship borne of those days was one with a fellow Texan, though no one thought of Jerry Mitchell as a Texan, and he certainly didn't encourage it. Jerry was everything that a stereotypical Texan was not—intellectual, urbane, nonacquisitive, and not interested in either oil or athletics (except for the cerebral sports like tennis and fencing). He was definitely born in the wrong state, probably on the wrong side of the Atlantic, and arguably in the wrong century. He had a distinctly continental outlook, classical values, and refined tastes to a degree that I seldom associated with his hometown of Dallas. Most remarkably, he had managed to pass into adulthood without the slightest hint of drawl in his speech, something that I had been unable (and had not tried too hard) to avoid.

Jerry was a graduate of Southern Methodist University, where he had started research with J. L. McCarthy, a student of the noted endocrinologist M. X. Zarrow at Purdue. Another of Zarrow's students, Jerome Yochim, had joined the faculty at Kansas, and that was the connection that had brought Jerry to Lawrence. Jerry had inherited from his mentors a compulsion for thoroughness, a talent for experimental design, the manual skill to bring it about, and the intellect to tie it all together. I was frankly in awe of him in the beginning, though after I discovered a sense of humor under the formal, controlled exterior he projected, I needled him mercilessly about Texas, football, money, and all the other things for which he claimed contempt. We hit it off well, above all because we shared the notion that science is a calling and a privilege and that literally nothing in life is more interesting.

In November of that first year, I met through Dorothy one of her fellow graduate students in philosophy, Bob Godbout and his wife Muriel. Born and brought up in the working-class neighborhoods of Manchester, New Hampshire, they met for the first time as seniors from different high schools at a church bazaar in the summer of 1960. Bob was a serious and committed scholar of the Thomistic philosophy curriculum at St. Anselm's College, and Muriel was working hard to help get her family through tough times. By Christmas of 1964, with Bob in his senior year at St. Anselm's, they had decided not only to marry but to break their bonds with New England. Bob knew that he wanted to teach at the college level and wanted the option to teach at a non-Catholic institution. The best way to do that, while satisfying the wanderlust he was feeling keenly by then, was to take a master's degree at a Midwestern state university of decent repute, then return to the East for the honing and polishing that he assumed the Ivy League could best bestow.

They were married in Manchester on August 21, 1965, and set out immediately on a honeymoon trip toward the University of Kansas in

Lawrence, half a continent away. That same night, I sat in the balcony of an auditorium in Lubbock, Texas, waiting for my name to be called to confirm that my straggling undergraduate years were over—oblivious, of course, that my life and theirs were about to converge.

The trip that started in hope and promise for Bob and Muriel turned into an ordeal, featuring a broken-down car, a midwestern thunderstorm that rolled at them in waves with a fierceness they had never experienced, and arrival in Lawrence well into the night, exhausted and tense, with no one to welcome them or show them the way to the married students' apartment they had reserved. When at last they found it and flipped on the light, half a dozen roaches bigger than any insect that ever grew in New Hampshire scurried for cover. It was the breaking point for Muriel at the end of a trying trip, and she cried the first night in her new home in an alien land.

With the dawning of a new day though, Muriel determined to make the best of her predicament. If she was far from her former home, she was at least and at last in a home of her own. In the days to come, she cleaned and organized and turned the two small rooms and tiny kitchen into a cozy if constricted habitation for the two of them—a home where all who came to know them felt warm and welcome and secure.

The confidence that the university had shown in providing Robert Godbout with a teaching assistantship was rewarded, as he quickly showed himself to be an outstanding scholar. His only peer was a Woodrow Wilson Fellow from the University of Texas, Dorothy Haecker, whose philosophical orientation had arisen, like his, out of the great and deep questions of religion. It was inevitable that they would be drawn to each other with all the approach and avoidance that characterize the friendship of top competitors. Bob and Muriel both yearned for new friends, and Dorothy could not survive without cultivating fairly intense relationships herself, so she was welcomed into their home. It was many weeks into the first semester of that first year in graduate school for all of us before circumstances brought about an introduction, through Dorothy, between Bob and Muriel and me.

We were instantly drawn to one another, fascinated by the factors that distinguished us. Bob, a Roman Catholic from New England, a philosopher of obvious intellect, looking at almost everything in metaphysical terms. I, a Texan of a religious persuasion they couldn't pin down, grounded in science, more amused than impressed by metaphysical issues. Unlike many of the scientists Bob had known, though, I did take philosophy seriously; and I had the great advantage of not being his classroom competitor. So we could argue philosophy and science and the interface between the two long and late on many evenings without ever threatening each other. Muriel's intelligence and insight were keen as well, but she tended to stay out of these rhetorical frays, content to either ignore Bob and me or regard us with bemusement from time to time. She had a seemingly insatiable tolerance for my corny humor and my considerable appetite for her cooking. There was nothing she prepared that I didn't like, and by telling her so frequently, I garnered dinner invitations at closely spaced intervals. Her generosity and compassion were boundless. As she came to know me better, her evident concern for my happiness was deeply touching and a tremendous boost to my ego.

The Godbouts and I came to rely on each other. I took them to shop, to do laundry, and to attend to the chores beyond walking distance from their apartment since their broken-down car had not been replaced. In return, they took upon themselves the responsibility, as much as I would let them, for my general well-being. They spent a lot of time explaining to me the rationale of Roman Catholic morality, to no avail, and I spent one afternoon teaching them to play poker, to their everlasting delight. In the beginning, we were a foursome—the Godbouts with Dorothy and me. After a time, our friendship developed independently of Dorothy, as she characteristically moved faster in a number of other directions. Increasingly, we spent our Sundays together, talking about science and philosophy, geography and politics, Texas and New Hampshire, and our very different backgrounds.

By April of 1966, they had come to grips with their new situation and wanted to venture out onto the plains to see for themselves the land they had known only from television and the movies. With a sense of adventure, which was palpable for them, we set out for Dodge City during spring vacation. When we had driven maybe two hundred miles, we came upon a railroad overpass, and they had me stop the car so they could get out and stare for a long time at a flatter earth and a farther horizon than any they had ever imagined.

Summer of '66

With Christmas approaching that first semester, Dorothy and I decided we would return home for the holidays through Houston, where a philosophy meeting that she wanted to attend was being held and where I might conceivably try to contact Georges Ungar. With uncharacteristic brashness, I called the Baylor College of Medicine and was put through to his lab. To my surprise, he answered the phone himself; and after I had expressed my interest in his research, he invited me to come in for a visit. The same cordiality I had inferred from his signature on the reprint he had sent me was on full display in person. After we had talked enough for him to realize that I knew a good bit about the current research on learning and memory, I asked if there would be any possibility of working in his lab during the coming summer. He said there might be and encouraged me to send him a formal application. I left his lab in a daze of disbelief over my good fortune.

The stopover in Houston also gave me a chance to catch up with Penny, who had returned to Texas Tech and was also home for the holidays. We had a happy reunion. Like two teenagers getting back together after separate summer camps, we could hardly talk to each other fast enough. Since this was Penny's city, she decided to show it off to me. She took me downtown, where a new wave of skyscrapers was just beginning to sprout, then out to the Medical Center, already impressive to me, but at a fraction of the mass it would become. Eventually, we found ourselves at The Athens, a bar for Greek sailors down at the Port of Houston waterfront. With a bottle of Greek wine in hand, and a fair amount of carousing in the background, we got down to serious conversation. We talked of the trauma in her personal life and the mess I had made of mine; of our numerous flaws, which, at the age of twenty-two, still loomed so large; but of the present excitement in our lives as well. Penny and I were close and had shared some of the happiest and the saddest events of our undergraduate years. It was unfortunate that during the roughest of those times, by reason of circumstance and hesitation, I had not been able to help her when she needed a friend nor confide in her when I needed one. But this was an occasion for mending our bridges and looking to the future. The affection and camaraderie between us were still there, and that realization alone made the memory of that evening linger fondly among my many recollections of Houston.

I finished my first semester in graduate school with the highest grade in my physiology course and was the top student in my experimental psychology lab (take that, Yale!). By February, a formal offer of appointment as a research trainee at the Baylor College of Medicine in Houston at $400 per month—a phenomenal sum to me then—arrived from Ungar. The spring semester ground on toward the summer, not nearly fast enough for me. In April, I was informed that Samson had decided to stay another year at NRP. This depressed me because it meant that another year would go by before I would get to solidify the type of relationship with my graduate school advisor that I had long anticipated. Any guilty feelings I had about going to Houston that summer certainly evaporated at that point.

On May 14, I was forced to take part in one of the more tragicomic consequences of the growing war in Vietnam. Thousands of male students gathered in auditoriums on college campuses across the nation to take a two-hour exam to determine who deserved to maintain their student deferments and who would be subject to the draft. The knowledge that the answers to those questions—a collection of general topics—could determine who would go on to pursue the usual ambitions of youth and who would die in a war far away against their will left a bitter taste of that day that I've never forgotten.

As I reached the sultry bayou of Houston the first week in June, I welcomed the warmth and humidity after what had been a much cooler spring in Kansas than I was used to. I settled in an apartment near the medical center, subleased from a friend of Dorothy's, and started to carry out experiments that Ungar and I had agreed to. We would use performance in a Y-maze as our learning task. Our donors would be rats trained to turn one way or another in a Y-maze to avoid being shocked. Our recipients would be mice untrained in the same maze, injected with brain extracts from the donors. If the mice showed a tendency to avoid the shock by running into the same arm of the Y-maze as their donors had been trained, transfer of memory would be demonstrated. Assuming that this could be achieved, we would go on to test whether extinction of learning in the donors would be accompanied by loss of the ability of memory to be transferred. Also, all our experiments would be conducted with extracts of brain tissue prepared in a way to contain proteins or peptides, but not nucleic acids.

As I got started on the thirteen- and fourteen-hour days that the experiment required, I also was given access to Ungar's extensive file on all the other experimenters around the world who were chasing the fame that a confirmed demonstration of molecular transfer of memory would represent. While many researchers were getting positive results, others were not. In particular, backlash against the claims of memory transfer by RNA had set in. A failure to replicate the report in August of 1965 by Jacobson's group at UCLA on transfer of approach to a food cup was published in *Science* by the end of the year.[3] *Science* published in 1966 two more failures to replicate the RNA transfer experiment by an aggregate of laboratories—the first by Jim McGaugh and his colleagues,[4] the second including Ed Bennett, Mark Rosenzweig, Robert Galambos, and Murray Jarvik, among others.[5] Though the procedures were significantly different in most cases from the method used by Jacobson's group, these negative results were particularly damaging because of the stature and established reputations of many of the coauthors.

Ungar and I were not overly concerned at this point by these negative results because they all focused on RNA, and McGaugh's group[4] had shown that little, if any, RNA gets into the brain when injected into the abdomen, as most transfer experimenters were doing. Furthermore, we believed it highly unlikely that RNA alone could affect neural function. Not only was there no plausible way we could think of for RNA to be decoded into meaningful neurophysiological activity, Ungar had shown that enzymes that break down peptides interfered with the transfer effect, while enzymes that break down RNA did not. This observation had also been reported by Frank Rosenblatt, who claimed more specifically that successful transfer of learning was mediated by polypeptides in the molecular weight range of 1,000 to 5,000.[6]

Among the early group of scientists seeking evidence for molecular transfer of learning, Rosenblatt was the most prolific and, by way of background, the most unique. He was a computer scientist, already famous as the creator of the Perceptron—the first computerized neural network capable of learning.[7] He was thoughtful and incisive in his experimental designs, thorough in his methodology, and open-minded about the possibility of information stored in chemical form in the brain, despite coming to the field from a tradition of strict connectionism. By the summer of 1966, he had already carried out a number of experiments that, in the aggregate, had persuaded him that the transfer phenomenon was real but complex—depending on a number of factors for which few investigators were controlling.

As my own experiments proceeded through the summer, I got initial results that strongly indicated a transfer of behavioral tendencies, but the more I considered data from various controls, the more I came to agree with Rosenblatt. By the time Ungar and I published our work the following year,[8] we had concluded that a number of critical factors affected the success of the transfer phenomenon, including stimulus modality, dose of the brain extracts, length of donor training, and recipient bias. No lab that we were aware of had reported a failure to replicate Ungar's results on transfer of morphine tolerance or sound habituation. So with Rosenblatt in agreement that, when transfer of learning occurred, it was most likely mediated by a small polypeptide, Ungar pressed on. A couple of floors below, Roger Guillemin was beginning to study the production of peptides in the hypothalamus—a region of the brain just above the pituitary gland—that controls a number of endocrine and other physiological functions. In time, he would be awarded the Nobel Prize for this work, and other discoveries would leave no doubt about the neuroactive properties of many specific peptides. But in 1966, this was still far from routine acceptance.

My summer was grinding on toward what I hoped would be a final, decisive experiment when the machinists' union called a strike against United, Eastern, and TWA airlines. Nothing that wasn't human or human luggage could get on an airplane for over a month, so my reliable supply of rodents stayed grounded in their home cages in New Jersey. Then calamity was visited upon the catastrophe of the airline strike when the air conditioner in my apartment broke down. When I got home at night, the apartment was an oven. I had to stay at work until midnight or spend my evenings in the tepid swimming pool to survive the heat wave that descended on the city. When I was too restless for either, I would drive out to Hobby Airport (Bush Intercontinental had not been built yet) and watch the non-struck airlines (Continental and Braniff—neither still in existence) land and takeoff, as if by so doing, I could will the slumbering Easterns and Uniteds and TWAs up into the sky and toward New Jersey. I owned one share of Continental Airlines, purchased for fifty-five hard-earned dollars a few months earlier, and I took genuine personal pride in every landing and takeoff of a "proud bird with the golden tail," as they were advertised then. I also greatly admired Braniff's outrageous campaign to do away with "plain planes" and loved to watch their gaudy orange and green and blue fuselages strut across the tarmac. It was an era of individuality, and I was all for it.

With extra time on my hands, I found myself contemplating my situation. In February, I had seen the movie version of Nikos Kazantzakis' *Zorba the Greek*. Anthony Quinn plays a man who deals with the darkest moments in his life by dancing on the beach. The main message of the story is that to be truly free in life, we have to have a little madness. It occurred to me that all my life, I had floated under control through the world of ideas, but I had never been crazy. I had never learned to dance. Now I was ready to dance on the beach, like Anthony Quinn. At the time, I drew no parallel between these personal reflections and what I was trying to do in my professional life. It is tempting to look back on the transfer experiments as my way of cutting the cord to the conventional science of the time—as a little dance on the beach in the early years of my career. But I didn't think of it that way then. The transfer experiments were a strain on established views, but they weren't beyond the pale of plausible science. Like Braniff, I had a fear of being plain, but I still wanted to fly like everybody else.

When the airline strike ended the second week in August, it was too late to carry out my last experiment. The summer was ending, and I had to head back to Kansas for my second year in graduate school. A heavy warm mist enveloped the city my last night in Houston, emblematic of the fate that my high hopes for the summer had suffered. But working with Ungar had been a pleasure and a privilege, and I had been eager and excited to face the morning nearly every day I was there. In time, the molecular mechanisms of memory would be known, I was sure, just not as soon as I had hoped. The rain fell harder as I headed north on the freeways I had learned to navigate with conviction, casting an eerie pall over the ghostly skyline of the city that had become an indelible part of my being.

For a night and a day, I drove north from Houston through a constant rain that spoiled the scenic route of my return through Arkansas, driven by my commitment to return to Lawrence despite the fact that it felt less like home than the city I had just left behind. On the afternoon of the second day, the sun came out as I descended onto the flatlands of western Missouri. When I crossed into Kansas, there was something about the site of the sunflowers, bathed in the light and waving in the early autumn breeze, that gave me a sudden sense of coming home after all. When the university buildings that tower over Lawrence came into view, the sense grew profound and unmistakable. I drove straight to the Godbouts, who were delighted to have me back. Muriel embraced me, Bob shook my hand, and we talked late into the night, that first night in Lawrence that I truly felt at home.

Surviving a Close Call

The next day, I located a comfortable if creaky apartment for sixty-five dollars a month. The next Friday, I woke up with a chill and swollen lymph glands. As the day progressed, I felt worse but had to work at registration all day as my left foot inexplicably grew painful and swollen. I mentioned this casually to Bob and Muriel at the end of the day, and they insisted I go to the clinic. I protested that it was nothing, but they prevailed. The nurse was appalled to discover a lower leg perilously infected from a torn toenail. She ordered me to bed immediately after a shot of penicillin, but again I protested, as I had a teaching assistants' meeting the following morning and, more importantly, the Kansas-Texas Tech football game the following night. The nurse was unyielding, but did agree to release me into Muriel's custody, whom she correctly perceived was the only person in Lawrence with sufficient moral persuasion to keep me at home. So the Godbouts took me to their apartment, where I suffered through a night of high fever and considerable pain. By morning, both the fever and pain had abated, and I assured Muriel that she had made much ado about nothing. But she still wouldn't let me go to the football game. Whether I would have had the common sense to go to the clinic in time to save my foot the previous evening, I don't know; but I have my doubts, and I never thanked Muriel enough.

[1]   Ungar, G. and C. Oceguera-Navarro. 1965. Transfer of habituation by material extracted from brain. *Nature* 207: 301–2.

[2]   The way in which Fred Samson and Frank Schmitt had met each other—and, in time, become highly codependent—is recounted in detail in Irwin, L. N. 2007. *Scotophobin: Darkness at the Dawn of the Search for Memory Molecules.* Lanham, MD: Hamilton Books.

[3]   Gross, C. G. and F. M. Carey. 1965. Transfer of learned response by RNA injection: failure of attempts to replicate. *Science* 150: 1749.

[4]   Luttges, M., T. Johnson, C. Buck, J. Holland, and J. McGaugh. 1966. An examination of "transfer of learning" by nucleic Acid. *Science* 151: 834–7.

[5]   Byrne, W. L., D. Samuel, E. L. Bennett, M. R. Rosenzweig, and E. Wasserman. 1966. Memory transfer. *Science* 153: 658–9.

[6]   Rosenblatt, F., J. T. Farrow, and S. Rhine. 1966. The transfer of learned behavior from trained to untrained rats by means of brain extracts. II. *Proc Natl Acad Sci USA* 55: 787–92.

[7]   Rosenblatt, F. 1958. The Perceptron: A probabilistic model for information storage and organization in the brain. *Psychol. Rev.* 65: 386–408.

[8]   Ungar, G. and L. N. Irwin. 1967. Transfer of acquired information by brain extracts. *Nature* 214: 453–5.

11

Synaptosomes, Gangliosides, and Chaotic Conferences

I had returned to Kansas from Houston with the notion that I would continue to work out the details of the transfer phenomenon as the next stage of my graduate research. Since Samson was going to stay away for another year at NRP, I wrote to him to share my thoughts and solicit his about plans for my doctoral research. He was not at all pleased when I told him I would like to follow up on my work in Houston. He was highly skeptical of the claims of biochemical transfer of learned behavior, by Ungar or anyone else. I pointed out that all the claims of failure to replicate the transfer experiments involved RNA rather than peptides, as Ungar and Rosenblatt were claiming, and that no failures to replicate Ungar's experiments had been reported. I also drew his attention to the growing evidence that the brain's hypothalamus secretes small peptides that regulate hormone release from the pituitary and therefore likely have neuroactive effects on brain cells as well. But Samson was unconvinced. He was vehemently opposed to Ungar, for reasons I would learn only much later,[1] and made it clear that he would not be supportive of my continuing that line of research. I saw the handwriting on the wall and wrote to him in November that I had decided to attack the problem of memory mechanisms in a different way, undoubtedly bringing him great relief.

Carol Lee Crumrine

Working at the registration desk, as I was the day my foot became infected, was a traditional assignment for upper-class graduate students that was definitely more of a duty than a privilege. However, it did give the majority of us who were males our first opportunity to assess the usually meager roster of incoming female graduate students. If they were at all appealing, we tried to sign them up for the lab sections of the first-year courses that we knew we would be teaching. The only woman to register out of four new graduate students that semester was Carol Crumrine, a chemistry major who had done her undergraduate work at Kansas. When she came to the desk, I applied all the Texas charm that I could conjure up discreetly, but it had no perceptible effect. She signed up for the Friday section of Mammalian Physiology, a section in which I would assist only for the first hour in order to attend a psychology seminar later in the afternoon. The possibility of social entanglements appeared bleak.

I did keep a casual eye on her during my one hour a week in her lab, however. She seemed quiet, shy, and a little tense. She was cordial to but not particularly impressed by me, except that she did seem grateful for my help with the small animal surgery at which I excelled. Another graduate student was in charge of that section, so I didn't have a chance to impress her with sustained eloquence or erudition.

As part of my determination to meld the graduate students of the department into a community of friends and scholars, I had arranged a social function in October that had to be cancelled for some reason, and this gave me an opportunity to call all the first-year graduate students. Since my excuse was ironclad and nothing much was at stake, I called Carol first. I never got to the others, because the conversation I had planned for three minutes lasted more than sixty. Unlike the lab, which seemed to restrain her, the phone left her open and apparently eager to talk. We learned a lot of the basics about each other and discovered what seemed to me more than a coincidental number of overlapping interests -- from travel, to German, to her hometown of Tulsa.

She had been born in Houston but raised in Tulsa, where she graduated from Will Rogers High School one year behind Anita Bryant, with whom she had little in common. She had also lived in California after her parents were divorced. She was a National Merit Scholar who probably could have gone to college anywhere. She wanted to go to Stanford since her boyfriend was at Cal Tech, but her father insisted on the University of Kansas, where he figured *he* had learned as much and as well as he would have on either coast. Just getting away from home was such a blessing that Lawrence turned out to be fine. She spent a summer in Germany and took advanced courses in the language, but inspired like so many of us by an outstanding chemistry teacher in high school, she was drawn to that subject and decided that it presented a greater career opportunity than German. Lacking a desire to enter the workforce as an industrial chemist, she decided almost in an offhand manner to go to graduate school. Since she knew the faculty and curriculum at Kansas, she chose to start there while she contemplated her longer-term future.

I ran into her a few days after our phone conversation, when one thing led to another, and I found myself invited to bring a copy of an old exam to her at her apartment just down the street from mine, where she had just moved in with a new roommate. The errand that evening turned into a two-hour visit. She and her roommate, Marilyn Hall, offered cake and coffee, as I had mistimed my arrival just before their evening meal; and this time, we got to the serious topics, like politics, religion, and organic chemistry. For me, it was a magical encounter, with a social chemistry among the three of us that couldn't be doubted. It left me in a daze as I walked slowly home then lay awake for a long time, alternately fighting and inviting the feeling that my life stood on the brink of a new and major development.

Carol and I had a number of other conversations after that first visit with her and Marilyn; then the social gathering I had planned originally was finally held. Art Friesen, my office mate, and I decided to invite Jerry Mitchell as an honorary first-year graduate student, even though he was a year ahead of us and definitely considered by us to be our intellectual and scientific superior. The party was held at my apartment and came off well. Eventually, it wound down to Art, Jerry, Carol, and me. Then Art left. Then Jerry left. Carol and I talked till four in the morning, peeling away gradually the increasing complexity that we were finding in each other. By the time we had our first formal date—to a German play—we had become more than casual friends.

In mid-November, Art and I had our first exam in cell regulatory mechanisms—one of the best courses I was to take at Kansas. But I felt that I hadn't done well on that first test (I was right), and I brooded about it most of the day. That night after studying for a bit, I stopped by Carol's office and asked her to go out for a late-night snack of doughnuts, after which we ended up again at my apartment, talking late into the night. It was clear by the end of that evening that our feelings for each other were running deeper. As I was returning home, having walked her back to her own apartment, I was amused by the reflection that the cell regulatory mechanisms exam, which had begun as the most important event of the day, seemed now so inconsequential.

Gangliosides

It was one thing to abandon the transfer approach, but another to come up with an alternative. I wish I could recall what first drew me to gangliosides. I have a vague memory of making a note the first time I read that there was a particular molecule localized, so it was thought at the time, in nervous tissue; but not a single textbook from a course I ever took bears such a notation, so my memory must derive from a desire to show that my insight at that earlier time was greater than it really was. Whatever occasioned my initial acquaintance with gangliosides, they were fully a part of my awareness by the time of my return from Houston. By then, I knew that they were thought not only to be concentrated in the nervous system but to occur in greatest density at the synaptic junction between nerve cells

Gangliosides—so named by the German scientist Ernst Klenk in 1942 because they were isolated from neural ganglia—are a combination of long hydrocarbon chains and a variable number of sugars and amino sugars, including at least one nine-carbon sugar called sialic acid (Fig. 3). Because the hydrocarbon chains are hydrophobic (soluble in lipids, but not in water),

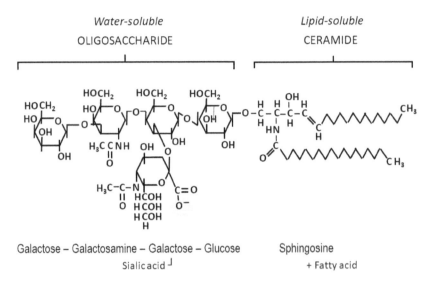

Fig. 3. Schematic structure of a ganglioside molecule. The squiggly lines to the right represent hydrocarbon tails which are embedded in the lipid core of the membrane, while the water-soluble hexagonal carbohydrate residues project into the extracellular space, where they can bind reversibly with cations, such as sodium and calcium, especially at the acidic carboxylic acid residues (shaded oval).

they are anchored in the lipid core of the plasma (outer) membrane of cells, while the polysaccharide (sugar and amino sugar) portion of the molecule is hydrophylic and extends into the aqueous space between cells. Under physiological conditions in the brain, the sialic acids expose a negatively charged oxygen atom that can bind reversibly with positively charged ions like sodium and calcium. This means that gangliosides could influence the flow of current in the extracellular space and possibly modulate the excitability of the cell. Their complex molecular shape could also provide binding sites for specific antibodies and other information-rich macromolecules.

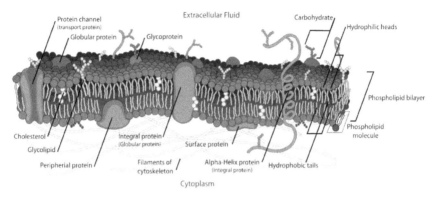

Fig. 4. Schematic diagram of cell membrane. A bilayer of lipids forms the membrane core, in which proteins, glycoproteins, and glycolipids (like gangliosides) are embedded, projecting their carbohydrate portions into the extracellular space, where they can interact with ions and other substances and influence the conduction of electrical currents. (Credit: LadyofHats [Mariana Ruiz], Wikimedia Commons.)

Synaptosomes

In 1962, a method was published for isolating the nerve endings from homogenized brain tissue.[2] Grinding up brain tissue under specified conditions could pinch off the nerve endings containing the synapse with its postsynaptic membrane still attached. This pinched-off end of a neuron resealed to form a spherical particle called a *synaptosome*. Gangliosides were found to be concentrated in synaptosomes, though they were also found in other subcellular fractions called *microsomes* in which membrane fragments of dendrites were concentrated. Dendrites form the extensive branches of neurons, which carry excitation mainly toward the nerve cell body for outward relay along the nerve cell's axon. Synaptosomes and microsomes differed in size and so could be separated by a centrifugation technique.

I began to reason that if gangliosides are involved in changes at the nerve ending that affect the strength of synaptic connections, as envisioned by connectionistic theories of neural plasticity, then analysis of changes in ganglioside composition in synaptosomes could provide evidence for such plasticity. This would definitely point to a molecular mechanism for memory as a *thing in a place*. Alternatively, shifts in the composition of gangliosides or other sialic acid–containing molecules, like glycoproteins, in dendritic membranes could alter the path and/ or intensity of current flow in the extracellular space, reflecting or affecting dispersed patterns of neuronal activity. These changes would be more likely to show up in the microsomal preparations of dendritic membranes and would be more compatible with a mechanism of neural plasticity that was a *process in a population*.

These experiments would be technically difficult to carry out, but they made logical sense, so I decided to write them up as a proposal for a National Science Foundation (NSF) predoctoral fellowship. NSF fellowships have always been very competitive and prestigious, and I didn't think I had much chance of winning one, but the possibility of having my tuition and a full-time stipend without any teaching duties paid for the remainder of graduate school was too attractive to pass up. So I brought my overall research plan into focus, proposing to become the first to provide definitive evidence that the synaptic connection between nerve cells is indeed altered when an animal learns.

On December 8, Samson arrived from Boston to meet with all the members of his lab and start writing a renewal application for his long-standing grant on energy metabolism in the brain. Finally, I was going to get to meet in person with the professor I had come to Kansas to work with fifteen months earlier. He had read my NSF Fellowship application on the plane and was very complimentary of it. He even proposed part of it for inclusion in the renewal application for his grant. Even though the rest of the group talked him out of it, the fact that he had included my thoughts in his research planning made me feel truly a member of his lab at last. This was part of a three-day meeting, characterized as usual by Samson's enthusiasm and creativity, accompanied by an excess of wandering speculations, nebulous objectives, and lack of organization. In subsequent years, I would come to experience this tortuous method for integrating and creating scientific ideas in the presence of Frank Schmitt, its master practitioner.

The next day, Samson and I finally had our one-on-one meeting. It went well, though I was disappointed that we didn't get to talk much about my plans for future research in the likely case that I didn't get the fellowship. But our student-mentor relationship was at least finally off the ground. Later that week, Carol and I went to the departmental Christmas party together. It was our first department-wide function as a couple, and I was quite pleased and proud to take her. We had a good time, tempered only when I learned with dismay that Samson had left town that morning before we had finished our talk about my future. I was going to go it alone a while longer, it seemed.

Germinating the Seeds of the Society for Neuroscience

As plans for my research program were crystallizing, movement was underway toward organizing that interdisciplinary society for bringing together all the researchers on the nervous system that the formation of IBRO in 1960 had set in motion. One of IBRO's early actions was to call for all its member nations to survey the activities and resources for conducting research on the nervous system in their respective countries.[3] In the United States, the National Academy of Sciences set up a Committee on Brain Sciences to do that and appointed Ralph Gerard to chair the committee.

Ralph Gerard had worked with A. V. Hill and Otto Meyerhoff, making important discoveries on the biochemical aspects of nerve conduction in the 1920s. He joined the faculty at the University of Chicago, his undergraduate alma mater, in 1928, where he continued to make important contributions on different aspects of neural function. In 1955, he was recruited to help establish the Mental Health Research Institute at the University of Michigan. He led that institution to a position of prominence, though failed in one of his earnest attempts: bringing Bernard Agranoff and James McConnell into a fruitful collaboration. In 1964, he was persuaded to move to the University of California at Irvine, where he helped establish one of the first interdisciplinary departments of neuroscience, the Department of Psychobiology.[4] Soon, similar departments would proliferate in other academic institutions throughout the United States and Europe and eventually the rest of the world, as neuroscience became a recognized field in its own right.

The Committee on Brain Sciences held periodic meetings from 1965 to 1969. By June of 1967, the committee had come to the conclusion that a single, overarching organization that unified all the various disciplines concerned with research on the nervous system would be better than the alternative then under consideration: a consortium of all regional special interest groups in existence at that time. So the committee appointed an "Executive Group for the Organization of Brain Sciences," to be chaired by Ed Perl of the University of Utah. The goal of the executive group was to create an interdisciplinary society of neuroscientists and to ensure its viability through the first few years of its existence. Perl drafted a constitution and bylaws for the new society, which was adopted by a twenty-member executive committee on June 16, 1969, at the National Academy of Sciences building in Washington. Those in attendance became founding members of the Society for Neuroscience.

Of critical importance, the founders favored an open and democratic form of governance, with liberal criteria for membership. The eventual procedure adopted was simple nomination by any two existing members, and the organizing committee was asked to propose neuroscientists to be invited to join without even going through those minimal requirements. As someone known to Frank Schmitt, I was among those to be automatically admitted to membership and so became the 223rd member of the Society for Neuroscience.

[1]   The mystery of Samson's antipathy toward Ungar, as well as my correspondence with my absent doctoral advisor throughout my first two years in graduate school, are detailed in Irwin, L. N. 2007 *Scotophobin: Darkness at the Dawn of the Search for Memory Molecules.* Landham, MD: Hamilton Books.

[2]   Gray, E. G. and V.P. Whittaker. 1962. The isolation of nerve endings from brain: an electron-microscopic study of cell fragments derived by homogenization and centrifugation. *J. Anat.* 96: 79–87.

[3]   Marshall, L. et al. 1996. Early history of IBRO. *Neuroscience* 72: 283–306.

[4]   Kety, S. S. 1982. Ralph Waldo Gerard, 1900–1974: A biographical memoir. Washington, DC: National Academy of Sciences.

12

New and Winding Roads

On the national scene, 1967 began ominously with President Johnson's State of the Union speech on January 10, proclaiming grim determination in Vietnam. It turned tragic on January 27, when a flash fire during a simulated launch for the first orbital Apollo flight killed astronauts Virgil Grissom, Edward White, and Roger Chaffee. I think there were many like me—teenagers when the first satellite was launched and drawn to science by the glamour and excitement of those early days when outer space crossed the boundary from fiction to reality—who took this setback personally and hard.

Ironically, the day of that tragic accident was the occasion of a small personal triumph for Art Friesen and me. It was the day of the final exam in cell regulatory mechanisms, and we each did well enough to make an A in the course. I knew that Art would make an A, but mine was a close thing and all the more rewarding. It had been a demanding course and was the grade of which I was most proud in all of graduate school.

A further irony came on February 6, when I learned I was one of about ten graduate students at the University of Kansas to be awarded a NASA Graduate Fellowship. The benefits, including full stipend with no teaching duties, were just as good as the NSF Fellowship would have been, so I would be free to devote full time to my coursework and research, whether I got the NSF Fellowship or not. What the NASA Fellowship lacked in prestige was compensated for by the symbolic importance of this award to me personally, coming so soon after the nation's first casualties directly related to the exploration of space.

On March 6, President Johnson announced his intention to issue an executive order calling for the drafting of the youngest men (nineteen-year-olds) by random lottery and the abolition of graduate student deferments. The charade of the previous spring, in which we had been tested to determine our worthiness for deferments, were now null and void, supplanted by a system of no deferments at all. I understood the argument that there should not be a privileged class exempted from harm on arbitrary grounds, and I knew that the draft fell heaviest (though by no means totally) on minorities and the less affluent; but I had not stumbled into graduate school by blind luck, nor had I survived five years of a rigorous liberal arts and science education without hardship. Graduate student deferments were not for arbitrary reasons. I truly believed I would be more valuable as a trained scientist than as a soldier in a war I didn't believe in. Nonetheless, it was hard to hear about the mounting casualties in Vietnam day after day without a rising sense of guilt and anger.

More on Gangliosides

"Hope all going well. I've thought about you a number of times since my visit to KU." So wrote Samson at the end of March to let me know I had not been forgotten. In responding, I decided to put in a plug for gangliosides again:

Evidence is growing that gangliosides are critical in synaptic function. It thus seems reasonable to me to find out whether or not gangliosides are changing when behavior is changing. If so, this would be direct evidence of changes in the biochemistry of synapses; a fact which has not been demonstrated to my knowledge, although it is widely assumed by memory theorists.

On his way to a meeting in Czechoslovakia a month later, Samson sent me a tip on extracting gangliosides that he had picked up from Leon Wolfe, an established neurochemist who "also thinks they are very important for brain function." I was already aware and making use of Wolfe's information but was pleased with this indication that Samson was sympathetic to what was developing into my major research interest.

Leon Wolfe's interest in gangliosides was an important factor, but not the only factor in Samson's growing attitude of acceptance. Samson was intrigued by Henry McIlwain's earlier work showing that gangliosides helped maintain the excitability of nerve cells studied in tissue slice preparations. At a pragmatic level, he gauged that there was some chance that work on gangliosides could lead to a discovery of real significance. As he once explained, "I figured that you could be very much on to it, and you could make me famous." Finally, at a philosophical level, he believed that his students did best at problems of their own choosing. Better to have me working on something I cared about with enthusiasm than going through the motions on a problem that fascinated him, but not me. Even if gangliosides didn't make either of us famous, little enough was known about them that the work would still be good basic research. And so it was that I came to work on an enigmatic class of molecules in the brain about which my doctoral advisor knew next to nothing. Later, when I was tempted to bemoan the lack of expertise he provided, I reminded myself that I couldn't have it both ways: I couldn't have a high degree of independence and a great deal of expertise all in the same lab. He let me make the choice, but the choice and the consequences were mine.

I was astounded and elated to learn on March 15 that I had been awarded the NSF Graduate Fellowship for which I had applied, contrary to my expectations. By that point, with the NASA Fellowship in hand, I didn't need it, but of course, I accepted it because of its greater prestige and selectivity. Only two years earlier, I had been rejected from graduate school at Yale, the University of Pennsylvania, and Western Reserve. As an undergraduate, I had come within one exam in one course of having to abandon my dream of a career in science. To now be accorded one of the highest honors the nation could bestow at my level of training was such a profound reversal of fortune that nothing was to bring me as much pride or satisfaction for the next ten years as that award.

More on Carol

Throughout the month of March 1967, as almost from the moment I met her, Carol's presence in my life and thought was pervasive. She had a measured, subtle demeanor and a quiet, confident personality that attracted me with an intensity that I honestly didn't understand. It was true that I wanted very much to remarry, but the healing process that had begun with Dorothy's help left me willing and able to stay single. Yet Carol seemed to embody so many of the virtues that I respected and admired, that not to think of marriage to her was like stopping the clouds from floating by.

I didn't dare tell her this because I feared she would be frightened if I pushed. For all our empathy and our evident pleasure at being together, her deepest feelings were an enigma to me. I got the sense that she cared for me, but as in all things, she let me know only in measured and subtle ways. From the first of the year, we had lunch and dinner often together, dated every weekend, and stayed up talking most of every weekend night. That she did all this willingly I took as a sign that she cared and that her affection could be cultivated. But I had the clear feeling that her affection, though growing, was a millennium or two behind mine. It imposed on me an emotional restraint that was taking its toll. How I got any research started that spring or managed to make grades that were commendable in my courses that semester is quite beyond my understanding in retrospect.

My involvement with Carol had some inevitable effects on my other friendships. My relationship with Bob and Muriel was the one most affected. I continued to see them often, but not as often as before. Bob and I still argued philosophy and religion frequently, but mostly now on Sunday afternoons, not Saturday nights. Muriel, who knew me well and worried about my happiness to a touching degree, understood the meaning of my less frequent visits and didn't begrudge them but, as late as January, had still not met Carol. I moved to mend this breach of protocol the first Sunday afternoon that Carol was back from the holidays. We only intended to stop at the Godbouts' for a few minutes, but ended up spending the afternoon there as Carol, in her unassuming way, won Muriel's enthusiastic endorsement.

Art and I spent less time together after our course in cell regulatory mechanisms ended, largely because we were spending more time with the women in our lives. He and his girlfriend, Marlyce—a medical student in Kansas City—planned a wedding for early July and honored me with an invitation to be an usher. The door to his dream was finally opened when he was admitted to medical school at Washington University. With medical school and marriage both imminent, he worked long, hard hours on the research for his master's degree. He probably did more research of higher quality in less time than anybody to come out of that department—certainly while I was there.

Jerry and I became closer friends. This was due in part to Marilyn, with whom Jerry had become infatuated, but their relationship was tempestuous in those early days and not a reliable point of common interest. Jerry had a way of answering questions he considered unduly probing with short, polite sentences that let me know he preferred to discuss the secretion of hormones or the latest exhibit at the Nelson Art Gallery rather than the status of his love life. And while he would never ask me anything so personal, he would tease me enough to let me know that he wasn't uninterested in the romantic entanglements of others.

My contact with Dorothy was waning by the time I met Carol, and after that, we saw each other quite seldom. Dorothy began to sing folksongs in the coffee cellars around the campus that were so popular at the time. Occasionally, I would go to hear her after a long evening of study or work in the lab. I would sit in the back, sometimes unknown to her, thinking about the nights years ago when she first taught me to play the ukulele. I also thought about all she had done for me and wished I could have done more for her. Our relationship was ending in the way that a road ends when it splits into two new paths that wind progressively away toward different destinations.

As the spring semester was ending, my relationship with Carol had reached a watershed. The spring had been lovely, like springtime in eastern Kansas can be, and our time together grew in contentment and serenity with the passing days. For the most part, we both worked hard during the week, living for the weekends that we shared. The more I learned of her troubled childhood, the more I marveled at the person she had become. While her lack of pretension remained constant, her confidence had grown as her graduate work had gone well, and under my influence (she claimed), her research had begun earlier than that of most of her colleagues. It was a delight to be with her, an incredible frustration to leave her. Though marriage had long been on my mind, it wasn't until the last weekend in April that I felt emboldened enough to ask her if she would consider the possibility of being married to me. Her response was warm and affectionate, but characteristically restrained. She was very honored to be asked and of course would consider it, but didn't feel that she was ready yet for a commitment of that magnitude. I was half prepared for this and wasn't as disappointed as I might have been, but the bottom line was a rejection—gentle and sweet, but a rejection nonetheless.

Having opened the subject, I had a hard time leaving it alone. For several successive weekends, I would raise the issue again, only to have it parried in a considerate but decisive manner. I began to dread the upcoming summer when, freed of coursework, we would probably be spending even more time together, with the only question that mattered in my mind being considered taboo. Finally, it dawned on me that she really didn't want to talk about it, as she kept saying. So in frustration, I decided that was fine: if she wanted to marry me, she would have to be the one to ask. We talked of marriage no more and had a great summer together.

Meanwhile, in Other News

Back in Houston, Ungar had adopted a new experimental protocol that was giving very robust results, based on an alleged successful

transfer of learned avoidance of the dark reported earlier that year.[1]
When given the option of entering a dark box or a lighted box, rats
invariably went first into the dark box; but if they were shocked as
soon as they entered the dark, they quickly learned to go to the lighted
box instead. Ungar had found that the dark-avoidance training, in
which rats were forced to be shocked initially in the dark box, was
highly reproducible. He also had found that he could transfer this fear
of the dark with peptide extracts quite well.[2] Encouraged by the ease
of training and the high reproducibility of dark-avoidance learning, a
number of researchers turned to this task as a means of trying once more
to probe for molecular residues of memory. This included Ed Bennett,
who remained skeptical; Bill Byrne,[3] who believed there was something
to it; and Arnold Golub, who had expected to find little but seemed to
be finding a great deal.

Arnold Golub grew up in California and enrolled in the University
of California, Santa Barbara, as a biology major. As he learned more
psychology, he decided to merge his two interests and become a
physiological psychologist. He was doing electrophysiological recordings
by his senior year with a professor who moved to Texas Christian
University (TCU) in Fort Worth, so Golub followed him there as
his graduate student. They eventually had a falling out though, so
Golub ended up working with Jim Dyal, a psychologist with little
biological or biochemical knowledge but who had heard about the
transfer experiments and was intrigued. Golub provided the biological
and biochemical expertise that enabled Dyal to try essentially a repeat of
Jacobson's transfer of learning by a rat to approach a food dispenser. But
Golub and Dyal changed the experiment in two important ways. First,
they retrained rats after their initial learning had been extinguished; and
secondly, they injected whole brain homogenates (which would have
contained peptides), not RNA extracts. Their result indicated a strong
transfer effect from retrained donors, but not from trained donors lacking
the extinction and retraining. This was yet another indication that the
transfer phenomenon was complicated with unresolved variables.

Research on memory disruption by antibiotic drugs continued with
vigor but increasingly was giving muddled results as well. The Flexners

discovered curiously that saline injected into the brain four to ten hours after puromycin injections a day after training blocked the amnesia induced by the puromycin.[4] They later discovered that puromycin generates the production of peptidyl-puromycin fragments, probably due to incomplete synthesis of full protein chains, and speculated that the amnesic effects of puromycin may be due to a failure of memory retrieval rather than the absence of memory storage.[5] Barondes, furthermore, found that puromycin causes occult seizures[6] and blocks protein less successfully than another antibiotic, cycloheximide, which did not cause amnesia.[7] It began to look like the amnesic effect of puromycin was due to disruptive factors other than inhibition of protein synthesis. With both puromycin and actinomycin D turning out to be poor drugs for the inhibition of protein synthesis in behavioral studies, Barondes turned to acetoxycycloheximide, which had fewer harmful side effects. He found that this drug inhibited retention of T-maze learning six but not three hours after being injected, suggesting that short-term memory (after three hours) does not require protein synthesis, but long-term memory (after six hours) does.[8] Agranoff also had started using this drug to block memory in goldfish, with similar results.[9] Both researchers were concluding that the long-term consolidation of memory requires the synthesis of new protein, but short-term memory does not.

While deeply engaged in these learning experiments with mice, Barondes started wondering what the newly required proteins were doing to make them essential for memory consolidation. Were they in fact changing the functional connectivity of nerve cells at the synapse, as Hebb had hypothesized? It had long been known that material synthesized in the cell body of a neuron moves to its axon ending by a process of axoplasmic flow. Barondes wanted to find out how fast this occurred, possibly to correlate the time course of memory consolidation with the speed at which potential modifications in synaptic proteins could occur. He was able to show that amino acid precursors are incorporated into protein in whole brain homogenates rapidly, but start to decline even as the precursors continue to accumulate in synaptosomes.[10] This suggested that proteins were indeed being synthesized in the cell body, then being transported with some delay to the synaptic terminals. This technique for studying axoplasmic flow was widely adopted and initiated a large body of research in many labs, which soon demonstrated that different proteins flow down the axon at different rates.[11]

Frank Schmitt, at the Neurosciences Research Program (NRP), became aware of this fertile area of research and decided that a work session should be devoted to the subject. Paul Weiss was the acknowledged pioneer in the field, but his forceful personality and contentious nature concerned Schmitt, who thought the younger and more amiable Barondes would be a better choice to lead the session. My supposed doctoral adviser, Fred Samson, was a scientist in residence at NRP when the planning for this session arose. He became deeply involved in its execution and found the subject so invigorating that he ultimately turned his own research toward the study of microtubules— the subcellular filaments that move materials (not just proteins, it turned out) along the dendritic and axonal extensions of all nerve cells.

When Samson returned to Kansas as a full-time faculty member in 1967, his lab took up the study of microtubules in earnest. I did not, but one of the experiments that Barondes had conducted did interest me greatly. He had used glucosamine, an amino sugar, as a precursor in addition to amino acids in order to measure the synthesis of glycoproteins—*proteins* with polysaccharides attached (by analogy with glycolipids like gangliosides, which are *lipids* with polysaccharides attached). What he found was that, unlike the protein portion of a glycoprotein, which is synthesized in a neuron's cell body, the glucosamine was incorporated into the polysaccharide portion at the nerve ending itself.[12] This immediately suggested a method for rapidly modifying the structure, hence possibly the function, of a macromolecule at the synaptic junction. This was very similar to what I was thinking about the possible involvement of gangliosides in neural plasticity, as hinted at in my NSF Fellowship application.

In another part of that application, I had proposed a simple-minded test of the extent to which nerve cells could recognize appropriate connections; namely, by mixing synaptosomes (pinched-off nerve endings) from different areas of the nervous system and measuring the extent to which they coalesced. Ungar was enthusiastic about this part of my NSF proposal because it mirrored what he thought was happening in the transfer of learning. If each cell has a unique chemical label, and if a specific behavioral experience is mediated by a specified assembly of nerve cells; then in effect, each behavioral experience has a unique chemical representation specific for that experience. Ungar proposed that nerve cells active in learning a particular behavioral option (like turning right or left in a T-maze) elaborate a peptide that marks the specific pathways involved. When homogenates of the donor brain found their way into the brain of a recipient, the newly elaborated peptide identified and sensitized the neural pathways that mediated the same information in the recipient.[13]

The prime example of neurospecific recognition already known was the precise connections formed during development between the eye and the brain in the vertebrate visual system. Retinal cells, which respond to stimuli from specific points in the visual field, send axons to connect with cells at precise locations in the brain. This "labeled line" concept of neural organization, which states that what is perceived from where, in the animal's perceptual field is interpreted by the brain according to the pathway through which—and destination to which—the information is delivered. That, in turn, requires precise recognition during development between incoming sensory neurons (such as retinal cells from the eye) with centrally located destination neurons (like cells in the optic tectum or visual cortex), where the nature and source of the stimulation is perceived. In a theoretical paper in 1966, Richard Roberts and Louis Flexner had proposed that cells from the retina hook up with cells in the visual cortex during development in a highly precise fashion because proteins on the surface of a nerve ending specify that cell uniquely, allowing the target cell to which it connects to recognize it individually.[14]

Barondes carried the Roberts and Flexner proposal a step further by proposing that neural identities could be specified by the polysaccharide portion of glycoproteins and glycolipids.[15] Not only could this chemical information be added quickly at the nerve ending, as his experiments had shown, but the variety of polysaccharide structures that could be created had the potential to generate considerable synaptic plasticity. Like

Barondes, I believed that the polysaccharide portion of glycoproteins and glycolipids might well play a critical role in specifying neural pathways. Unknown to him at the time, my ideas about neurospecificity and plasticity were falling into synchrony with his to a remarkable degree.

In summary then, Ungar was suggesting that specific neural pathways are specified by a unique set of peptide secretions; Roberts and Flexner assumed that proteins play that role, while Barondes and I were suggesting that polysaccharides attached to glycoproteins or glycolipids provide the specificity of neural identities. Whatever the details turned out to be, the extent to which nerve cell surfaces are chemically unique and distinguishable had become an issue of overriding importance.

Meanwhile, Back in Kansas

The summer wore on in Lawrence. I built a maze for my rats, taught myself to extract gangliosides, mastered the assay for sialic acid, and completed a revision of the neuroendocrine section of a review chapter I had agreed to write with Ungar. Art and Marlyce were married in a lovely ceremony in Nebraska the night of July 1. I finished reading Ernest Hilgard's classic *Theories of Learning,* which argues in essence that the evolution of higher brain function was a consequence of the evolution of the vertebrate kidney.

Carol and Marilyn also worked hard and, by August, decided that they deserved a vacation. Having lived frugally their first year in graduate school and having had their social lives subsidized by Jerry and me, they had actually managed to save enough money from their $200-a-month stipends to take a trip to New York. With regret, I drove them to the airport in Kansas City where they took off for the East Coast, excited by the prospect of a grand adventure (though miffed a bit, in Marilyn's case, by the fact that Jerry was too busy to see her off).

By that time, my maze was finished, so in Carol's absence, I buried myself in training rats and looking for chemical changes in their brains. My first experiment was simply to train rats to turn either to the left or the right in a T-maze to avoid a mild electrical shock. The rats were

killed after a week of training, and their brains were fractionated to yield
pinched-off nerve endings (synaptosomes). The objective was to obtain
evidence for chemical changes at synaptic junctions. I measured total
protein, as a general class of molecule, and sialic acid, as a characteristic
component of gangliosides and glycoproteins. After long days in the
lab, I let myself into Carol and Marilyn's apartment to watch late-night
television (which I didn't own), while I calculated the data from the
day's work with a slide rule—an archaic mechanical instrument faintly
similar to an abacus.

I missed Carol terribly, but got more work done those two last weeks
in August than ever before or possibly ever since. I had planned a trip to
Texas at the end of the summer, not expecting to see Carol till the second
week in September. But on the morning of September 3, I was awakened
by a phone call from Carol in Tulsa, inviting me to come through
there on my way to Houston. She was back from her grand adventure
and apparently anxious to see me. Such an uncharacteristically bold
initiative on her part could only mean something good. By dinnertime,
I was in Tulsa. We had a happy reunion, constrained as it was in the
home of her father. She was full of talk about her trip and said nothing
about our relationship other than that she had thought about me a lot
while she was gone. Something about the way she said it conveyed ever
so slightly more feeling than similar assertions before, but this time, I
forced myself merely to let the words linger.

We spent the next day in Tulsa together—the day I was supposed
to be driving to Houston. So the following morning, I had to leave at
the crack of dawn, as the Ungars were expecting me for dinner. I made
it and enjoyed greatly my reunion with the three of them—Georges,
Alberte, and Catherine (their daughter, who was still at home).

Ungar and I worked all the following day on last-minute editorial
adjustments to our chapter. To my great disappointment, he had cut out
the entire neuroendocrine section I had written, and to my later regret,
I didn't challenge him to explain his reasons for doing so. While I was
reasonably pleased with the final product, I thought that its impact
was lessened by deletion of the growing information about neuroactive
peptides, not only because I thought it central to our overall point

about sensitivity of nerve cells to multidimensional chemical influences, but because I felt it represented the best chance of linking the transfer experiments to a plausible view of brain function. Time to do the latter was clearly running out. Ungar was always open to my criticisms and may well have been persuaded to leave the neuroendocrine section in had I insisted. But I didn't. It was the only time I didn't stand my ground with him, and the one time I really should have.

By the time I returned to Lawrence, both Carol and Marilyn were back as well. We spent a lot of time together for several days, as they had many adventures and anecdotes to relate. They delighted especially in recounting the afternoon that a couple of producers from the *NBC Nightly News* solicited their company in a bar near Rockefeller Center. They emphasized to Jerry and me that they spurned the advances of these two executives, but he and I noted with interest that they avidly scanned the credits at the end of the NBC news program for months thereafter.

Jerry was a trial to Marilyn. She had come to care for him a great deal, as he apparently did for her; but he was not prepared to have their relationship escalate into anything exuberant. It wasn't his way. The mixed signals he sent her left her confused. It was not unusual, therefore, for her to show up at my door as Carol and I would be having dinner or studying afterward. Knowing where she could find us both, and confidant that we would be sympathetic, she would vent her frustration over the missing member of our foursome. Occasionally, it would work in the other direction. If Carol were gone and I were feeling lonesome or unsure of my progress with her, I would go see Marilyn, knowing that I would come away with sympathy and reassurance.

Muriel kept a discreet but watchful eye over all this. Her top priority was to cope with Bob's metaphysical ups and downs, which seemed to be considerable, though I have a hard time these many years later separating the memory of my intellectual tribulations from his. Muriel was wonderfully tolerant of our endless esoteric conversations and incredibly supportive of her husband and all his friends, but I sensed that she wondered if it really mattered that much more than getting the electric bill paid on time.

The change in Carol's tone that I thought I had detected in Tulsa persisted as the fall of 1967 got underway. Something had happened on her trip to New York and New England, something that relaxed her caution and amplified her feelings toward me. On the fourth Friday night in September, we got to talking about our relationship again, and she volunteered that we might really "have what it takes" after all. This enigmatic statement left me an opening too big to pass up. I reiterated my desire to marry her and said that the offer was still there. She said that she had been giving it some thought, but didn't say anything more. After a long pause, she started talking about something entirely different. I couldn't believe it! I asked her if that was all she had to say on the subject of marriage.

"I guess so."

My level of frustration was escalating. It dawned on me that I had been too imprecise, so I decided to phrase the question directly.

"Would you marry me?"

"Yes."

"When?"

"I don't know."

"Christmas?" I pushed my luck.

"OK."

Triskaidekaphobia

In order to become a candidate for a doctoral degree in my department at the University of Kansas in 1967, a student had to demonstrate a reading knowledge of two foreign languages, pass a series of six short written exams on any subject the professor giving the exam fancied ("Why do leaves change color?" was my favorite), and pass a comprehensive oral exam administered by the student's doctoral advisor. By midway through my second year, I had passed the reading exams in French and German and passed my sixth written exam, so Samson let me schedule my oral exam for the fall of my third year, 1967. He suggested the date of October 13, which happened to be a Friday.

While I assured him I was not superstitious, neither did I see the need to take unnecessary chances and suggested that a week later would still fit in to both our schedules nicely. This only fueled his insistence that the most important exam of my whole career definitely be taken on Friday, the 13th of October.

As the bad day approached, I was confident but increasingly nervous. I stayed up till two thirty the morning of the exam, hoping to absorb just the small bit more that might be needed to put me over the top. At 10:15 a.m., we assembled in the conference room and began. My first inquisitor asked me for evidence of neural specificity. Incredibly, I gave a difficult and ambiguous example first, before recalling the simpler and more straightforward example of retinotectal connectivity. We moved to thermoregulation in lizards and mammals. I was strong on this and picked up steam. My physiological psychology professor asked me the physiological control for hunger in rats—a topic right out of the course I had just taken with him. Now I was rolling.

We then got into the subject of my research and spent most of the rest of the exam on that, which naturally worked to my advantage. We talked a good bit about the transfer experiments, and Samson took the opportunity to openly question the integrity of some of those doing them. I tried to shrug this off by saying the data were a matter of public record. To question their interpretation was scientifically legitimate, but to question their authenticity called for an ethical judgment that I wasn't in a position to make. It was a lively discussion, by the end of which my initial nervousness had certainly been dissipated. Jerome Yochim, the department's endocrinologist, ended the session by asking me the function of the hypothalamus. With the peptide releasing factors from that region of the brain still on my mind, as well as the conviction that they must represent a broader phenomenon than we were yet appreciating, the question was a cream puff easily disposed of.

It was all over with by noon. By a quarter past noon, I had been informed that I had passed (though not with honors; I never passed anything with honors). By 1:30 p.m., I was stretched out on the couch in Carol's apartment, watching the World Series between the Red Sox and the Cardinals, with a backache of mysterious but surely psychosomatic origin that grew to immobilizing intensity in the days that followed.

[1]   Gay, R. and A. Raphelson. 1967. "Transfer of learning" by injection of brain RNA: A replication. *Psychon Sci* 8: 369–370.

[2]   Ungar, G., L. Galvan, and R. H. Clark. 1968. Chemical transfer of learned fear. *Nature* 217: 1259–61.

[3]   Bill Byrne was the first author of the paper in *Science* the previous year that had reported with twenty-two colleagues using four different types of learning a failure to find any evidence of transfer by RNA. This placed a damper on the memory-transfer paradigm from which it never recovered (even though none of the experiments had attempted transfer with peptides). Ironically, over the next few years, Byrne became one of the most persistent advocates for the reality of memory transfer.

[4]   Flexner, J. B. and L. B. Flexner. 1967. Restoration of expression of memory lost after treatment with puromycin. *Proc Natl Acad Sci USA* 57: 1651–4.

[5]   Flexner, L. B. and J. B. Flexner. 1968. Studies on memory: the long survival of peptidyl-puromycin in mouse brain. *Proc Natl Acad Sci USA* 60: 923–7.

[6]   Cohen, H. D. and S. H. Barondes. 1967. Puromycin effect on memory may be due to occult seizures. *Science* 157: 333–4.

[7]   Barondes, S. H. and H. D. Cohen. 1967. Comparative effects of cycloheximide and puromycin on cerebral protein synthesis and consolidation of memory in mice. *Brain Res* 4: 44-51.

[8]   Barondes, S. H. and H. D. Cohen. 1968. Memory impairment after subcutaneous injection of acetoxycycloheximide. *Science* 160: 556–7.

[9]   Agranoff, B. W., R. E. Davis, and J. J. Brink. 1966. Chemical studies on memory fixation in goldfish. *Brain Res* 1: 303–9.

[10]   Barondes, S. H. 1964. Delayed appearance of labeled protein in isolated nerve endings and axoplasmic flow. *Science* 146: 779–81; Barondes, S. H. 1966. On the site of synthesis of the mitochondrial protein of nerve endings. *J Neurochem* 13: 721–7.

[11]   Ochs, S. and J. Johnson. 1969. Fast and slow phases of axoplasmic flow in ventral root nerve fibres. *J Neurochem* 16: 845–53.

[12]   Barondes, S. H. 1968. Incorporation of radioactive glucosamine into macromolecules at nerve endings. *J Neurochem* 15: 699–706.

[13]   Ungar, G. 1968. Molecular mechanisms in learning. *Perspect Biol Med* 11: 217–32.

[14] Roberts, R. B. and L. B. Flexner. 1966. A model for the development of retina-cortex connections. *Am Sci* 54: 174–83.

[15] Barondes, S. H. 1970. Brain glycomacromolecules and interneuronal recognition. In *The Neurosciences: Second Study Program*, edited by F. O. Schmitt. New York: The Rockefeller University Press.

13

Test Tubes, Techniques, and a Wedding

The Cardinals won the World Series of 1967, breaking the hearts of Boston fans again. With my move to Massachusetts still a decade away, I rooted for the Cardinals and lay in bed following my oral prelim exam, waiting for my back to get better. Bob Godbout was not worldly-wise about a lot of things at the age of twenty-five, but he knew all about backs, and he was certain that only a chiropractor could cure my ills. Having played a part in saving my life, or at least my foot, just one year earlier, he must have taken my rejection of his sound advice as arrogant, not to say ungrateful. But I did reject his advice and, fortunately this time, did get better without the intervention of a medical practitioner of any kind.

With my oral exam behind me, and with Carol won over at long last, I was in a buoyant mood—anxious to get on with the search for memory molecules. My first attempt at detecting chemical changes at the synapse associated with learning had yielded negative results, but I had a better experiment in mind and intended to get it started as soon as I cleared up some technical problems with the assay for sialic acid. This was the state of my mind as Dianna Redburn and I were getting to know each other.

Featured Vocalist at the Red Dog Saloon

Dianna Ammons was the second-born of five children to a locally prominent family in the northwest Louisiana town of Many (rhymes with *canny*). Her older brother taught her enough of what he learned in kindergarten to get her into the first grade by the age of four. Precocious in school, she came to be known for other peculiarities. When she was twelve, she decided she wanted to become a race car driver. In the eighth grade, she had a pixie haircut when the other girls wore their hair in a bouffant. And when it was a scandal to do so, she was the first in her community to wear Bermuda shorts in public.

With age, her individuality matured. By the time she left for college, she was rejecting the blatant racism of her community and becoming involved in the awakening civil rights movement of the South. Her father exerted a dominating presence in his community and his family. His expectations were high and demanding. Her mother was adaptable and accepting, with an ability to focus on the opportunities rather than the disadvantages of any situation. Their daughter much admired this ability that her mother possessed and always tried to put it into practice in her own life. As an adolescent though, the daughter was more like the father. But she rebelled against his dominance, his interests, and some of his values, though not his belief in education. She decided to go to college, but would go away to Centenary College, eighty miles to the north in Shreveport, where she found herself "three galaxies away from Many."

Liberated by the distance from the source of her rebellion and influenced by Mary Waters, chair of the Biology Department—"a very rigorous scientist and a very demanding mentor/teacher"—she was drawn to science. Nurtured by the guidance of Dr. Waters, Dianna applied for and received her first scientific grant, which she used to build an ultramicrotome. She made the first ultrathin sections of tissue at Centenary College and took them to Houston, 250 miles away, where an oil company owned one of the early high-tech instruments, an electron microscope. While Dr. Waters was renowned for her successful placement of students in medical schools, her star pupil never showed interest in the practice of medicine, so graduate school at LSU in Baton Rouge was the logical alternative.

Prior to the start of her graduate studies, Dianna received her second grant—an NSF Summer Fellowship to study marine biology at Gulfport, Mississippi. But there, during the summer of 1964, she began to get an inkling that marine biology might not be for her. It was also where her professional career on the stage flourished briefly. In the tradition of belles throughout the South, she had participated in beauty contests from an early age and had toured with the Centenary Choir, including a nine-week stand at Radio City Music Hall three summers previously (the obvious source of the rumor among her later colleagues that she had been a Rockette), so public stages were not a novelty to her. Still, she was surprised, after auditioning on a whim, to get the job of featured female vocalist at the Red Dog Saloon in Gulfport. The management liked her talent, but found Dianna Ammons from Many, Louisiana, lacking in pizzazz. She therefore became Diane Drew from Las Vegas and nightly poured her soul into songs like "My Heart Belongs to Daddy" to supplement her meager stipend from NSF.

Ray Redburn from West Plains, Missouri, with service in the navy but no college behind him, was working as a doodlebugger in the north central part of the state at the time. A doodlebugger is a member of a seismographic crew who walks around with antennas and other electronic gear strapped to his body, rather resembling a creepy, crawly arthropod of the same name. He rented a room from a lady in Homer, near the Arkansas border, who happened to be Dianna's grandmother. So impressed with Ray Redburn was Dianna's grandmother that she summoned her granddaughter for a visit. In that way, the doodlebugger from Missouri met the marine biologist and vocalist from either Louisiana or Las Vegas, he wasn't sure which.

Ray was at least as impressive as her grandmother had led Dianna to believe, and within a year, she was prepared to drop out of graduate school to marry him and be with him in his wandering occupation. If her grandmother was delighted, her parents were not at all pleased. They had expected a man of more substance for their daughter, someone whose work wasn't named for an insect. In frustration and petulance, her father's final advice was "Whatever you do, you get your teacher's certificate, because if this guy runs out on you, you can at least teach!"

She did get her teacher's certificate and taught as a substitute in the towns across the Gulf Coast oil basin, from Elsa to Edcouch to DeRidder. They moved six times during the first year of their marriage. But Ray did not run out on her, and he was not lacking in substance. He harbored ambitions of becoming an architect and enrolled in the Missouri School of Mines at Rolla, with the expectation that Missouri would open its first school of architecture there. Within the year, however, that had fallen through. Dianna taught science at the junior high school in Rolla long enough to know that it wasn't the profession for her. Dually disappointed, the couple looked to move again. The state of Missouri had an exchange program with the state of Kansas for academic programs offered by one state but not the other. With an architecture school of fine reputation at the University of Kansas, Ray and Dianna headed for Lawrence in the fall of 1967.

Dianna had arranged by correspondence to work for a limnologist at the university, but he wrote her at the beginning of the summer that he was seriously ill, and by September, he had died. With no better plan in mind, Dianna started walking from one science department to another, passing out her résumé, asking for a job of any type. As she was reciting her situation for the secretary in the Comparative Biochemistry and Physiology office, the chairman, Fred Samson, walked in to pick up his mail. Overhearing his secretary in the process of politely dismissing the applicant, he intervened spontaneously and without explanation. Dianna remembers it all in vivid detail.

"After she had said, 'I'm sorry,' whatever, and I was turning to go—I mean, it was all over—Fred said, 'Well, listen, if you've got a minute, come in' . . . That changed my life." Impressed that a faculty member would spend any time with her at all in the first place, she caught Samson in one of his usual expansive and enthusiastic moods. "He was very receptive and . . . charismatic, talking about neuroscience and the frontier. There was no turning back after the first fifteen minutes."

Everyone seems to remember their first encounter with Fred Samson. Dianna remembers it, she believes, because the commitment of those who come to love their work in science is as irrational as falling in love itself. You can't help remembering those first encounters, and those moments of decision that transform an interest into a passion.

Not Enough Shaking Going On

Fred Samson introduced me to Dianna, his new technician, in mid-September, telling me that she would be doing miscellaneous chores in the lab. *Great,* I thought, *now we can get caught up on our dish washing.* Unglamorous as it sounds, a lot of the labor in a biochemistry lab is washing glassware. The type of work that I and other members of the lab were doing at the time generated a lot of dirty dishes. Sometimes there would be well over a hundred test tubes by the end of the day, and they all had to be washed by hand—scrubbed ten times with a brush and soapy water, rinsed till the soap was gone plus ten times more, then rinsed ten times with distilled water.

Dianna was so grateful to find a job in a lab that she didn't know any better. She washed test tubes by the hundreds and cleaned animal cages to the point of nausea. But on the whole, she felt blessed, and soon, she showed herself adept at more than washing test tubes. With Art Friesen gone, I was the senior graduate student left in the lab. Operationally in charge of the whole lab was Dennis Dahl.[1] He was a research associate on Samson's grant, with the job of executing the overall research program. Samson was never in the lab; in fact, I never saw him do an experiment in my life. His administrative duties kept him in the departmental office. But he tried to keep tabs on all of us and was continually throwing out ideas and trying to keep us stirred up. I was a problem for him in the sense that my research on gangliosides, glycoproteins, and behavior were outside his historical line of work; but he was just as supportive, intellectually and financially, of my experiments as he was of the ones on his major grant. Thus, I thought nothing of appropriating Dianna for my own purposes, though technically, she was assigned to Dennis.

Dianna and I hit it off well. Our shared southern roots, our similar ages, and our broad biological backgrounds contributed, but our mutual enthusiasm for what we were doing was the main thing we had in common. We had, as Dianna put it, "probably the best job in the world . . . the luxury of playing mind games with nature. What could be more fun?" The fact that Carol and I constituted a couple opened the opportunity for a social relationship as well, and Dianna capitalized on this by inviting us out to their house trailer on the northern edge of Lawrence for dinner, in honor of the engagement that Carol and I had recently announced. The menu was southern fried chicken, which solidified our social contract.

At work, Terry Hexum and I were advancing Dianna up from dish washer as fast as we could. He had her doing enzyme assays, and I showed her how to do the sialic acid assays. Dennis sat in his office, trying to digest the torrent of ideas from Samson and devise a line of research of his own, so Terry and I had Dianna occupied daily. We tended to be less reflective than Dennis, more inclined to plunge into a procedure and work out the bugs through experience, instead of taking the time to anticipate the problems. It was less efficient, but more productive than Dennis's way.

Soon after I had taught myself to do the assay for sialic acid, I had begun to be bothered by its inconsistency. Triplicate samples (three separate samples of the same solution) differed in apparent sialic acid concentration much more than they should have, indicating some uncontrolled error in the reaction or, more likely, in my methodology. Dianna, in learning the assay procedure from me, learned whatever it was that I was doing wrong because she, too, could not get consistent results. Though the measurement of sialic acid was important to my line of research, my more central focus was the synapse and whether it changed chemically with learning. If it turned out that I couldn't measure gangliosides with accuracy, I would simply focus on another chemical end point. So one morning with reluctance, I came to the lab resolved to drop the sialic acid assays, hence gangliosides and glycoproteins, out of my research program if I failed to get the sialic acid assay to work one more time. I would let Dianna make the final try, as I knew she was as good at the assay as I was; but I didn't tell her that it was then or never.

I was working at my desk when she came up to me, beaming self-satisfaction, with the news that she knew what was wrong with the assay. We weren't shaking the test tube hard enough! The last step of the procedure involved extracting the rose-colored indicator into an upper layer of organic solvents from the lower aqueous layer where the chemical reaction takes place. I had always done this by up-ending the capped test tubes ten times—exactly ten—in leisurely succession. She repeated this procedure for me, the way I, hence she, had been doing it. Then she took another sample, scrunched her eyes shut, and shook it as fast and as hard as she could, fifteen or twenty times in rapid-fire succession. Sure enough, after the layers separated in the tube that had the living daylights shaken out of it, all the color had moved into the upper layer, while in the tube that got more genteel treatment, traces of the color remained in the lower phase. Whether she reasoned this out, tried it at random, or discovered it by accident, I never knew; but I realized instantly that she was right. It was a trivial technical detail that no amount of coursework would have taught. But in the end, it was everything because it cleared away completely the inconsistency in the assay. I decided to stick with gangliosides after all, and did so for the next thirty years.

Back on Track

After the sialic acid episode, Dianna and I were on a roll. I was doing a lot of basic biochemistry with the assay at the time. It was labor-intensive groundwork that had to be done before I could move on to the interesting and important learning experiments. There were assays to run nearly every day, so Dianna was becoming invaluable.

I should have known it wouldn't last. While gangliosides and learning were the world to me, they were being bootlegged on a grant for the study of adenosine triphosphatase. Dennis cast an increasingly covetous eye toward the effort that Dianna was expending on my behalf and finally announced that, henceforth, Dianna would work on the subject matter of the grant from which she was being paid. I could have appealed to Samson over Dennis's head, but Dennis was in charge of the lab, and I respected his position, if not his productivity. Nonetheless, I didn't like it. As time went on, and I watched Dianna talking more to Dennis and experimenting with him less, I resented it. Dianna was being wasted while my work was piling up. But at least the dishes were getting done.

With Samson's return from Boston, he and I decided together on another experiment. We would test the involvement of gangliosides in neural function, without regard to their subcellular localization, by a more sensitive measure than I was using in my maze studies. We would measure the metabolic turnover, rather than just the concentration, of the compounds. The term *turnover* refers to the fact that molecules in a biological system are continually being manufactured (synthesized) and broken down. If the rate of synthesis is equal to the rate of breakdown, the molecules are said to be in a steady state, and their concentration remains constant, just as the number of people in a room stays the same if one person arrives for every one that leaves. This is presumed to be the usual situation. Note, however, that the concentration will also stay constant if the rate of synthesis speeds up (or slows down), as long as the rate of breakdown speeds up (or slows down) accordingly. Thus, even if the concentration does stay the same (as it must on balance if we are not to expand or shrink in size—our brains don't grow larger because we learn), changes in the turnover rate of a molecule may mean that it is susceptible and possibly involved in the cell's adjustment to whatever stimuli or other influences impinge upon the cell.

Our choice of behavioral manipulation was based on maximizing our chances of bringing about a detectable change in the biochemical endpoint. We used as our model the environmental enrichment approach that Rosenzweig, Bennett, and their colleagues had been using for years at Berkeley. In our version, rats in the stimulated group were housed together, exercised daily, forced to swim, exposed to loud noises and bright lights, and given plenty of opportunity to move about and explore the environment. Rats in the isolated group were housed individually and placed in a dark room with plenty of food and water so that they didn't have to be disturbed at all for a week. Notwithstanding the fact that the dark wasn't such a good idea (since rats are nocturnal and more active at night), the experimental design was probably effective in providing a lot more information for the nervous system of stimulated rats to deal with than the brains of their isolated controls.

By mid-November the experiment was up and running, and the initial results were promising. It looked like ganglioside turnover was significantly greater in the stimulated than the isolated rats. Gangliosides were being put into and taken out of the neural mosaic much faster in brains that had to cope with a lot of information than with a little. It was the first link between gangliosides and information processing—extremely tenuous but very exciting.

Time Out for the Wedding

The long days of training rats, extracting gangliosides, and doing sialic acid assays required by these experiments unfortunately were not relieved by evenings of domestic relaxation. While I approached my upcoming marriage to Carol with a sense of peace and equanimity, Carol was beset by growing apprehension and turmoil largely beyond her control. She assured me that she wasn't having doubts about me, but the whole matter had aggravated the delicate relationship between her divorced parents and set her mother into a state of agitation that had Carol and other friends of her family worried. At the suggestion of one of them, Carol began to consider the possibility of moving our wedding up to Thanksgiving.

We began to look for apartments with urgency, and after a futile search for something more economical, we settled on a pleasant but overpriced unit at $105 a month. We ordered a nice set of rings; but soon after collecting our $85, the company went out of business. They promised that our rings would be delivered, but not possibly before Christmas. The situation with Carol's mother worsened, so Carol decided to go for the Thanksgiving wedding, despite the fact that one of her intended bridesmaids would not be able to attend. Bob Godbout, of course, would be my best man.

I was still running experiments on Tuesday before Thanksgiving, the last hectic morning before leaving for Tulsa. We arrived there too late for blood tests but early enough to go through the ordeal of informing both of Carol's parents independently of our decision to get married on Friday. Carol's father took it pretty well. Carol's mother did not. Wednesday, we had our blood tests, acquired our license, and arranged with a skeptical minister to perform the ceremony two days hence. Toward the end of Thanksgiving Day, I got a call from Bob and Muriel with the regrettable but not unprecedented news that their car had broken down in southeastern Kansas, and there was no way they could make it to Tulsa for the wedding.

At that point, then, it was the night before our wedding; and we were down one set of rings, one bridesmaid, one best man, a set of parents that were absent (mine couldn't come), a set of parents that were not happy, and a bride- and groom-to-be for whom the wedding was turning into something of a catastrophe. We worked at containing the disaster. Marilyn, who *had* managed to make it to Tulsa from Lawrence, kindly consented to stand in place of the missing bridesmaid. And a friend of Carol from childhood, Richard E. Johnson, whom I had met less than twenty-four hours earlier, agreed to stand with me, a total stranger, as I married one of the women to whom he felt closest in the world.

The day dawned crisp and bright in Tulsa the morning after Thanksgiving. Perhaps the wedding would come off well after all. For me, it did; I rather enjoyed it. For Carol, it was not to be. The hairdresser did an awful job in the morning. Her mother was still in a twit. There were last-minute hitches. By three o'clock in the afternoon, she was walking down the aisle in a daze, living for the moment when the thing would be done and we could get out of town. I was so proud to be married to her and pleased to meet her strange and delightful circle of friends that I wanted to linger at the reception. She, on the other hand, was not in a mood to dally. We excused ourselves after one piece of cake and retreated to an undisclosed destination. Through it all, she seemed to hold me blameless and professed contentment at being alone with me at last. The next day, we drove back to Lawrence. She had a class on Monday morning, and I had rats to run.

Fig. 5. Louis Irwin and Carol Crumrine on the eve of their wedding on November 27, 1967.

Moving On

Just before Christmas 1967, Georges Ungar wrote to congratulate Carol and me on the news of our marriage and wish us happiness. He also reported great progress with his newfound paradigm of dark-avoidance training. His ability to transfer avoidance of entry into the black box was so reproducible in his lab that he was thinking of using it as a bioassay for proceeding to the next step: isolation and chemical characterization of the molecules responsible for the apparent influence of the behavior of naïve recipients.

I thoroughly endorsed this approach; it was the next logical step for a scientist whose international reputation had been built on his successful use of bioassays. My own trajectory by then was in a different direction. Ungar thought my experiments on gangliosides, synaptosomes, and so forth were too mundane; but he didn't have to deal with Samson, who considered the transfer studies as something barely short of fraud. I didn't believe *that*, having gotten enough positive results with the transfer paradigm myself to believe that peptides were affecting behavior in some way. But I *had* come to the conclusion that a straightforward equivalence between a molecule and a memory was almost surely wrong and that the search for memory molecules, as such, was misguided at best. But with most researchers in the field, I recognized that some degree of chemical change had to happen somewhere in the brain for acquired information to be retained. Finding what those chemical changes were and where they might occur may have been a long-shot with the methodologies in use at the time, but no science course I had ever taken had taught me not to take chances. And I didn't see anybody doing anything better.

My personal life was becalmed at last—a state that I relished. Symbolic of this was the way the year ended. John Giele, Carol's fencing coach through her championship undergraduate years (she was Kansas State champion in women's fencing in 1963) had become my friend as well as hers and began to visit us frequently soon after we were married. One of his first visits to our new apartment was on New Year's Eve. The three of us had a quiet, sedate, but very enjoyable private party to issue in the new near. I was grateful to see 1967 end, confident that the coming year would be considerably less tumultuous.

I was so wrong.

---

[1] The critical role that Dennis Dahl and his wife, Nancy, played in Samson's research program and my own progress through graduate school is recounted in detail in my previous book, so will not be repeated here. See Irwin, L.N. 2007. Scotophobin: Darkness at the Dawn of the Search for Memory Molecules. Landham, MD: Hamilton Books, pp. 47-49.

14

Things Fall Apart

For my generation, no year can match 1968 for turmoil on the national scene. By the time the year ended, the United States would be witness to two political assassinations; a drug culture gone awry; a widespread military victory in Vietnam perceived as a defeat that precipitated, ultimately, a real retreat; a president with the greatest popular mandate in history precluded from running for reelection, while a man that nobody seemed to like much took his place; and above all, the exacerbation of divisions along every conceivable line—those of age, race, and class being only the most obvious.

Death and Despair

The year was young when we were first touched by sadness. Bob called on January 2 to tell us that his brother, Richard, was missing in action in Vietnam. On January 3, he called to report that Richard had been killed near the Cambodian border the previous weekend. On January 4, I drove Bob and Muriel to the airport in Kansas City—a long, quiet drive—so they could fly back to New Hampshire to grieving parents and Richard's wife, whom he had married a week before his departure for Southeast Asia.

On January 17, President Johnson had nothing new to say in his State of the Union speech. His policy on Vietnam was unyielding; our will was being tested severely; our duty was grim perseverance. Everything about Vietnam was depressing. North Vietnamese and Vietcong forces launched a broad and massive military offensive on January 30 during the holiday season of Tet. By the end of February, it was a military catastrophe for North Vietnam and the southern insurgents. But in this country, so used to optimistic pronouncements about the progress we were making toward winning the war, the scope and intensity of the offensive had a shocking effect. The American public began to realize that it had been misled.

On Sunday night, March 31, President Johnson spoke to the nation about the war. He announced an unconditional halt to the bombing of most of North Vietnam. It was the boldest initiative for peace he had ever taken. But the president went on to shock us and the rest of the world by announcing at the end of the speech that he would not accept his party's nomination for reelection. While he couched it in terms of the need to pursue all avenues for peace unfettered by political considerations, it was in effect a stunning admission that the people no longer supported the position he represented. For that reason, to me, his announcement was great news.

Very soon—within minutes, it seemed—Marilyn was in our apartment, as upset as I had ever seen her about anything. She was genuinely distraught that our president, the first one we all had voted for, was being driven from office. In my pleasure at Johnson's implicit admission of defeat, I had not been sensitive to the full symbolism of the event. Marilyn, an intensely partisan Democrat, saw the magnitude of the fall of this towering political figure in more personal and classically tragic terms. Carol and I both tried to put the best face on it, offering the interpretation that his withdrawal was an act of statesmanship and courage. The reflection that Marilyn forced upon us, however, muted our celebration and made us focus in sympathy and respect on the meaning of Lyndon Johnson in the development of our own political awareness: his mobilization of national grief into action following John Kennedy's assassination; his conversion of the civil rights movement into national policy; his well-meaning, if thwarted, war on poverty and urban decay; even his understandable, if flawed, perception of the meaning of political insurrection in the developing nations of the world, conditioned as it was for his generation by the dreadful, overlearned lessons of the Second World War.

For just a few days, Johnson basked in the glow of good feelings that his twin decisions to restrain the bombing and withdraw from the presidential race had engendered. Then came a senseless act that signaled the onset of a series of national tragedies that would leave us all remembering the year with a grimace. Martin Luther King, in the midst of supporting a sanitation workers' strike in Memphis, was assassinated. This set off another summer of urban violence and mayhem across the nation.

With our depression lingering for weeks, Carol and I made the unprecedented decision to go see a movie on a Tuesday night. An unknown actor, Dustin Hoffman, was playing in a Mike Nichols movie downtown that interested us mildly, so we went to see *The Graduate*. Only a handful of movies have ever moved me profoundly. I have mentioned already the importance of *Zorba the Greek* to the development of my own self-understanding. *Dr. Zhivago* touched me deeply when I was first becoming romantically involved with Carol. *The Collector* I found lovely and haunting. But *The Graduate* blew me away. Were I to see it today, I think I might wonder why it touched such a nerve. I believe it was because there was so much promise in 1968, yet so much frustration—so much to be happy about personally and professionally, yet so many reasons for sadness in the system of which we were a part— that the stunning ending of *The Graduate* perfectly encapsulated our anxiety. With the possible exception of *Alice's Restaurant*, no movie lives in my memory of that time like *The Graduate*.

Jim McGaugh came to the University of Kansas on the first of May. I was excited. I was going to meet another legend, and we were going to exchange ideas on mechanisms of learning and memory. His seminar was excellent, as I had predicted; his pharmacological dissection of the processes of memory encoding and retrieval were as eloquent in their oral exposition as they were in the published record of his work. So my anticipation of the meeting with him and Samson the following day was heightened.

We sat in the same conference room where I had passed my doctoral oral exam eight months earlier, but this day, the outcome would be memorable only for what wasn't accomplished. What *did* transpire was a lot of talk about administrative trivia and academic biopolitics between Samson and McGaugh. There was some gossip about mutual acquaintances, though McGaugh was not a great fan of the NRP, so that would have limited the range of subject personalities. What did *not* get discussed, because there was no time left after the two department chairmen had rambled for the entire allotted time over matters of no consequence, was a critique of the current spate of memory theories or an analysis of whether the synapse was the only reasonable locus for the molecular residue of memory or whether glycoconjugates like gangliosides were reasonable candidates for informational macromolecules in the brain. I blame myself more than them. In later years I probably frustrated students of my own in the same way, at some time or another. I should have been more assertive. I should have been more insistent. My mother had tried to instill it in me. But it wasn't in my nature.

Robert Kennedy, on the other hand, was assertive and insistent and rolling across the political landscape of the nation, having entered the Democratic contest after Johnson had shown vulnerability when Senator Eugene McCarthy finished strongly in the New Hampshire primary. Vice-President Hubert Humphrey, who entered the race after Johnson's withdrawal, supported the administration's position on the war and had the backing of the party regulars. On Tuesday, June 4, Kennedy showed that he was probably the only anti-war candidate with a chance to beat Humphrey by winning simultaneously in South Dakota and California. At midnight, he gave his victory speech in the ballroom of the Ambassador Hotel in Los Angeles, then took a shortcut through the kitchen to a press conference that he never reached, because a man in the passageway shot him for a cause unrelated to the presidential campaign.

Carol and I had watched the election returns on television but gone to bed before the tragic ending of the day on the West Coast. We awoke the next morning to the grim news that Kennedy was gravely wounded. We didn't feel like working, so we went through the motions of the day, running errands and preparing for a small celebration in the evening, as it was Carol's birthday. Marilyn came for dinner at Carol's invitation, and John Giele dropped by in obvious distress. We began to drink and tried to be merry, stubbornly intent on staying jovial in the face of the impending tragedy. At 1:44 a.m. on June 6, Kennedy's inevitable death was officially declared. For most of the evening, the four of us had managed to laugh together, but by two o'clock in the morning, there had been too much alcohol and too little happiness in our party. Carol and I got sick, then went to bed.

Stranger in a Foreign Land

Earlier in the year, Carol and I had engineered compromises with our respective doctoral advisers. To my grave disappointment, Samson once again was lured by Frank Schmitt back to the NRP for the summer. Knowing that I hoped to finish my research within the year, he felt guilty enough to ask Schmitt if I could join NRP for the summer, where we could at least talk. Schmitt agreed, as having me there at no salary was apparently a small enough price to pay to get Samson back. Carol was well underway with her own research by then, and absence for a whole summer from the lab of Dick Himes was not something she was even willing to ask for. Plus, we had only been married for a few months at that point and had no intention of being separated that long so soon. So I proposed to Samson that I join him at NRP for just two months, and Carol got Dick Himes to let her be gone just for the month of August.

So it was that I drove alone to Boston the first week in July. At times of confusion in unfamiliar places, I gravitate toward airports; and since I arrived in Boston in confusion late in the day on July 6, 1968, I ended up at Logan International. Nothing in my previous experience had indicated that it would be foolhardy for me to try to find my way around a new city—like all southwesterners, I prided myself on my sense of direction and ability to read a map. But Boston was not part of my previous experience, and after I had wandered for a couple of hours throughout the metropolitan area, totally lost and disoriented in a futile search for MIT or just a motel that I could afford, I found myself heading through a tunnel toward the airport. There, the familiar looking planes with their well-known colors and logos brought back a sense of cultural identity and soothed me enough to search again, with success this time, for a motel where I could rest and collect my wits.

In my attempt to find Cambridge the next day, I wandered into the area of the Harvard Medical School and there discovered the Countway Medical Library. Knowing that an excellent library nearby was bound to be an asset for whatever I ended up doing for the coming two months, I decided to search there for an apartment. In time, I found a small room with a hot plate and half a refrigerator in a tenement building on Huntington Avenue. A young African American woman, whose only name I ever knew was Sandy, watched me move in; and when she heard me speak, she tagged me as a southerner (Texans sound like southerners to northerners) and put her guard up. Fortunately for both of us, her curiosity overcame her caution. She struck up a conversation that turned into a two-hour dialogue and, ultimately, a fond but ephemeral friendship. By definition, all southerners were ignorant and bigoted in her view. By the end of our first meeting, I think I had disabused her of that, but I felt that in place of ignorance, I had conveyed—accurately, I'm sure—a tremendous sense of naivete.

My third day in Boston, I finally located Samson and the NRP at the Brandegee Estate in Brookline. A nook was found for me somewhere in the bowels of the edifice, in what must have been part of the servants' quarters in a more august age. I was introduced at long last to Francis O. Schmitt, the founder and director of NRP; to Gardner Quarton, Program Director; Ted Melnechuk, Director of Communications, resident poet laureate, and most gifted thinker; and George Adelman, the librarian who lacked the healthy ego of most who worked and passed through there and who became my closest friend at NRP.

My assignment was to look over material collected in a work session many months earlier on aspects of the brain cell environment. The organizers had failed to come through with the promised written summaries, and the staff writers at NRP didn't have enough technical knowledge to produce a summary on their own, so Schmitt and Adelman had just about given up on it. But the session was right down my alley, since it dealt with the biochemical nature of brain cell surfaces and the conductivity of the intercellular space into which proteins, glycoproteins, and gangliosides project from those cell surfaces. Any changes in the nature or density of those cell surface molecules could alter the conductivity of the intercellular space, as Adey had long been saying, or affect the excitability of brain cell membranes, as everyone agreed. Not only did I have the knowledge to absorb and integrate the information presented at the work session, but it bore directly on my own doctoral research, so I was happy to be given the chore of seeing what I could salvage from the three-day meeting.

On July 14, the first weekend after my arrival in Boston, George and Sandy Adelman took me to Revere Beach. It was there that I first came to grips in a fundamental way with the culture of the northeast. It was a lousy beach day by any standard I had ever known. The temperature hovered in the high sixties or low seventies, and a fog hung over the coast. Yet there on the beach were thousands of people, stretched out as far as the eye could see. On a day like that, not a soul would have been caught dead on the beach at Corpus Christi or Galveston, and even on a nice day the beach would not have had that kind of crowd anywhere on the Gulf Coast (though today, the same could not be said). The sheer density of the population in the East, then, was my first realization; and the second was that the northeasterner's concept of a nice summer day was decidedly more liberal than my own. Starvation for sun was the only way I could explain it, and I took pity on them. In return, someone took advantage of me. It hadn't occurred to me to lock the door after I changed and left my clothes and wallet on the back seat. Upon my return from the sunless chilly beach, my wallet was gone. Sandy Adelman was sorry for my loss but appalled at my ignorance upon discovering the unlocked door. It amused her greatly to tell of this incident for years thereafter.

My pity on the sunless northeast was premature, as the following week saw a heat wave descend that was as uncomfortable as any in Kansas or Texas. On most evenings, I escaped the oven-like atmosphere of my tenement by spending several hours at the air-conditioned Countway Library, the most fabulous collection of scientific literature I had ever encountered. On fairly frequent occasions, before or after treks to the library or sometimes in lieu of them, my tenement friend Sandy and I would engage in long talks about our backgrounds and the current ills of society. She enlightened me to the nuances of the Black Power movement and taught me about the diversity of the black community, its ideologies, and its aspirations. I sympathized with all this intellectually before I met her, but she provided me with first-hand experience of the reality. In me, she learned to recognize the limitations of her own stereotypes. Terrible though the war in Vietnam was, the two consecutive summers of racial unrest in this country had served, especially after the death of Martin Luther King, to preoccupy the minds of caring people with this topic even more than the war. At the height of the nation's racial anxiety, Sandy and I came to know each other simply as two interesting people, rather than as a northern black and a southern white. If nothing else had been accomplished, that would have made my time in Boston worthwhile.

On the last three days of July, I had my first experience with that grandest of NRP functions: a Stated Meeting of the Associates (the invisible college of NRP). Paul Weiss, Robert Galambos, Neal Miller, Ted Bullock—names I knew well from the literature—were there in the flesh for me to meet for the first time. The high point of the meeting for me was a long conversation with Paul Weiss, perhaps the most dominant figure in developmental neurobiology in the twentieth century. By then in his seventies, he was bright, animated, and seemingly interested genuinely in my questions and comments. I had written of his work fairly extensively in the chapter with Ungar, and to meet him now and talk to him as a colleague was quite a treat.

Others I met that I knew less well but was greatly impressed by included Eric Kandel and Gerald Edelman. Kandel that year published with Alden Spencer a comprehensive review of the state of research on cellular approaches to the neurophysiological mechanisms of learning.[1] They pointed to the relevance of developmental biology for neural specificity by noting that "some specific influence not directly related to the transmission of electrical potentials is imparted by nerve cells to the structures they innervate." And with regard to molecular substrates of memory, they insinuated as highly unlikely the assumption that hereditary and experiential information would be stored in the same way. Furthermore, "At this early stage . . . it may be premature to emphasize the role of any single molecular species, or of any specific biochemical mechanism in neural plasticity," they wrote. That same year, Edelman published over a dozen papers, including five on the covalent structure of a human immunoglobulin in the journal *Biochemistry* alone.

The Elusive Chesterfield Hotel

By this time, Carol had arrived. She was appalled at our housing, but the time was short so we decided to hunker down in our one-room apartment with hot plate and half refrigerator. From the moment she arrived, Carol was determined that I should see New York City at the first opportunity. As our bus entered the city through the Bronx, I was captivated by the incredible density of buildings and people. Then when we crossed into the upper reaches of Harlem and proceeded down through the brick-lined canyons of Manhattan toward the Port Authority Bus Terminal, I found myself at a loss for adequate adjectives. I was awed to silence by the massiveness of the city and by its evident vitality on that hot summer evening.

Carol showed me how to negotiate the subway system. We emerged from a hole in the ground somewhere in mid-Manhattan near our intended destination, the Chesterfield Hotel. We walked several blocks, I carrying the luggage. We walked several more blocks, and Carol got a puzzled look on her face. It was right there, she assured me, pointing to a parking lot. Indeed it had been there the previous summer, the scene of her memorable stay with Marilyn. In the meantime, it obviously had been transformed to a parking lot. We walked back down the blocks we had walked up, I carrying the luggage. Finally, we happened on to the Taft Hotel, where I was prepared to procure a room at any price.

After settling in, we hit the streets. We strolled to Times Square, Rockefeller Center, up Fifth Avenue, down Madison, and along the incredible interface of city and parkland at Central Park South. We were up early the next morning, eating on the run at a Calico Kitchen, which left Carol queasy the rest of the day, but there was too much to see and do to be slowed by physical discomfort. We visited the UN, took in the Bowery, and toured Greenwich Village to see some real live hippies. The hippies intrigued me, the opulence of Rockefeller Center amazed me, and the derelicts of the Bowery disturbed me. I had never seen such wealth nor such poverty, certainly not so closely apposed and so brazenly taken for granted. Poor was something I knew; in South Texas, I had been in homes with dirt floors. But poverty on display, as a stop on a tour bus, was a new experience.

In the evening, we ate outdoors at Rockefeller Center's Promenade Cafe, then took in a show at that epitome of tourist traps, Radio City Music Hall. We loved it. To end the evening in proper style, we retired to the top of the NBC building for a nightcap in the Rainbow Room. Sitting there in the sky, the twinkling lights of Manhattan wrapping around and stretching away from us, I had the feeling of being in a rarefied atmosphere where quite possibly I didn't belong. It had barely been thirty-six months since I had bought my last beer at the Elbow Room in Kermit, Texas; but the distance in space, time, and culture from the Elbow Room to the Rainbow Room seemed unfathomable that evening, and I wasn't at all sure that I had really made the transition.

Back in Boston, my work on resurrecting "Aspects of the Brain Cell Environment" resumed as the parade of interesting people through NRP continued. John Smythies, an accomplished pharmacologist with a vivacious, engaging wife who, along with Sandy Adelman, was the hit of every NRP social function she attended, was one such visiting scientist. One day, Smythies gave a talk on *his* theory of learning—yet another one of the dozens at the time propounding a synaptic mechanism. It struck me as reasonable and useful, but I summoned my courage to ask if anyone in the room could cite definitive evidence that synaptic changes were the basis for learning. The answer, in essence, was no, but no one seemed to think that it was a particularly relevant question. From that day on, I began to suspect that brilliance had its blind spots, and a seed of cynicism toward NRP was sown in my mind. Meanwhile, to the surprise of Schmitt and Samson, and to George Adelman's delight, I produced a manuscript by the end of August that evolved into one of NRP's more influential publications.[2]

Back in the Provinces

Our sojourn in Boston thus completed, Carol and I headed for home. As I began the experiments that would culminate my doctoral dissertation research, the tragedy-laden presidential campaign was staggering toward Election Day with the field narrowed to three candidates: Hubert Humphrey, Richard Nixon, and George Wallace. Humphrey had been nominated the night of August 28, 1968, at the Democratic National Convention in Chicago while bedlam reigned on the streets and tear gas wafted through the hotels of the delegates. He paid a price for seeming too eager to accept the nomination of the party regulars and too insensitive to the excesses of the police riot in Chicago that coincided with his moment of triumph. In their bitterness, the college students who had fueled the campaigns of McCarthy and Kennedy by and large fell into a cynical and depressed silence. Humphrey's campaign languished while Nixon and Wallace capitalized on the nation's paranoia with their "law and order" rhetoric. I was disgusted with the events in Chicago and sorry that Humphrey had beaten McCarthy, but the alternatives (Nixon or Wallace) appalled me; so I did put in a few hours of campaign work at the student union on Humphrey's behalf, along with Marilyn, whose organizational efforts for the Democrats maintained some degree of visibility for an alternative to Nixon and Wallace at the University of Kansas that fall.

The turbulence of the times made those of us working on seemingly esoteric problems in research labs, especially those of us threatened by the specter of a very different type of activity in Vietnam, become introspective about the balance between the isolationism of the scholar and the public involvement of the citizen engaged in the social and political turmoil sweeping our society. In a long memorable discussion over dinner one evening that October, Carol, Jerry, and I struggled with the issue of whether a scientist could afford the luxury of keeping above the fray. Our natural preference, like that of most scientists, was an affinity for the ivory tower; but Jerry insisted that detachment was no longer a possibility, and we agreed. It portended the pattern of our later lives: Jerry would eventually become very involved in a number of civic affairs, and I finally undertook a foray into politics. But for many of the most successful scientists, scrupulous commitment to the ivory tower approach has been essential, and can we really say that on balance humankind is the worse for it?

Hubert Humphrey for too long remained immobilized by his loyalty to the Johnson administration's position on the war but finally managed to chart an independent course in a speech the night of September 30 by announcing that he would stop the bombing of North Vietnam if elected president. Then, confronted at last with the realization that the United States was not and probably could not bring the war to a conclusion by military means, President Johnson himself ordered a complete bombing halt on the last day of October, four days before the election.

There was a national sigh of relief and a surge toward Humphrey, but by a hair, it was too little too late. The nation with a 43.4 percent minority of the voters chose Richard Nixon to be its president—an act that would see the war in Southeast Asia prolonged and expanded for five more years and the first president of the United States resign in disgrace after committing criminal acts and lying about them to a public that by their vote had asked for it.

Cold, rainy weather for a week after the election added to my depression for several days, but I eventually threw myself into my experiments, and that perked up my spirits. I was doing it all on my own by then, as Dianna was assigned totally to other projects. But we worked in close quarters, and inasmuch as she had political views that were strong and mostly in the opposite direction from mine, we had spent a good bit of time arguing in a manner that was friendly but spirited enough to leave me testy on more than one occasion during the campaign. After the election, Dianna kindly refrained from gloating, and both of us started getting more work done again.

Marilyn's spirits were rising also, as her relationship with Jerry was improving. Jerry received word in late November that he had won a prized postdoctoral appointment at the Baylor College of Medicine in Houston. As the sole surviving son of a family that had suffered a casualty in Vietnam, Bob was able to pursue his graduate work freed from the dread of the draft. This relaxed him and relieved us all of one other small source of worry.

Carol and I paused to visit our families during the holidays. On Christmas Eve, Apollo astronauts Frank Borman, Jim Lovell, and Bill Anders became the first humans to orbit another world. The next morning as they circled the moon, they read passages from the book of Genesis, which offended a few ideological purists with no sense of poetry or cosmic perspective. The beauty of this accomplishment—as a technological triumph, a mission of peace, and an act of goodwill for the whole human race—contrasted so sharply with the conflicts and disparities on Earth that it was painful to contemplate. For me, that very contrast was what made the space effort so tremendously important and, in my view, showed its critics to be people of such limited vision.

Dash to the Finish Line

By the end of January 1969, my search for chemical changes in the dendritic surface as well as nerve ending fraction in rats trained in a one-trial swim-escape task I had designed[3] was beginning to yield exciting results. In short, they indicated that the surfaces of the neuron's dendritic branches were just as likely to change chemically in behaviorally stimulated rats as their nerve endings.

My final experiment was completed on February 26. This left me seventy-two days to write a doctoral dissertation and prepare for my final oral defense, if I wanted to graduate in June. On Easter Sunday, April 6, I wrote the first words of my doctoral dissertation, beginning with the purely descriptive sections on methodology, which are the easiest to write. The ensuing days of concentrated writing fell into a rhythm encompassing some new and unusual activities. To break the tedium of the desk work, I would go to the gym at noon and play handball with Dick Himes, Carol's doctoral advisor, and a couple of other graduate students in the department. The physical activity provided a good counterpoint to my mental exertions. A different counterpoint was provided by lunchtime breaks, during which I fell into the habit of watching soap operas. My interest was not casual. I fell in love with the character of Leslie Bauer on *The Guiding Light* and—this is the truth—was depressed for several days following her sad demise. I determined that if I ever had a daughter, she would be a Leslie. *As the World Turns* eventually won my loyalty as well, and I continued to enjoy the comedic genius of Eileen Fulton (Lisa) for decades thereafter.

As the days wound down to a precious few, I finished the experimental sections of the dissertation and started on the all-important introduction. As much as anything, my doctoral work would point to a new way of looking at research on chemical correlates of memory, so the introduction was critical. The major points were simple: (1) theories of how memory is encoded were a dime a dozen and not particularly creative—nearly everyone was looking for proteins or nucleic acids as the residue of memory and the synaptic junction between nerve cells as its locus; (2) few researchers were analyzing chemical changes with enough anatomical or subcellular detail to match the high degree of functional specialization in the brain; and (3) memory could not be readily dissected out of the behavioral context and internal physiological state of the organism within which it is encoded and retrieved. What was needed was a higher ratio of experimentation to theorization, a search for memory-related molecules other than proteins and nucleic acids at more localized anatomical levels, and the use of a variety of behavioral and physiological manipulations to which memory contributed in varying degrees.

As to my own research, I cited the rationale basically formulated at the NRP work session on the "Brain Cell Microenvironment," which I had written the previous summer; namely, that the cell surface and its immediately adjacent microenvironment constitute a promising focus for research on information processing in neural networks. For that reason, molecules characteristic of the cell surface, such as gangliosides and glycoproteins, deserved closer examination. My research represented an initial attempt toward that end. My experimental results were tentative and weak but were pointing the way to what I anticipated would eventually provide fruitful revelations about how the brain works at the molecular level.

I gave the first sections of the handwritten draft to the typist eight days before my final oral defense and started drawing the graphs and typing the tables myself. With four days to go, I gave the first typed sections to my dissertation committee to read. With three days to go, I finished writing the concluding section of the dissertation and got the final typed pages to my committee with two days left. At 5:30 p.m., the day before my oral defense, Jerome Yochim finished reading it and signed the form saying that I could proceed to my oral defense, the last member of my committee to do so. Carol and the typist worked late into the night making typographical corrections and a few changes requested by my committee, while I prepared for my final presentation and defense.

The next morning, I rehearsed my talk a couple of times, then went downtown to buy a new white shirt for the occasion. At three thirty that afternoon, I began the oral defense of my dissertation, "Biochemical Changes in Stimulated Brain and Mechanisms of Information Processing." About forty-five minutes later, I began to answer questions, and twenty minutes after that, it was all over. My committee congratulated me, as did friends and fellow students. Carol and I walked home together, exhausted but pleased, and not quite sure how so much had been accomplished in so little time.

I slept on the couch as the sun went down on May 9, 1969, totally absorbed in my personal accomplishment, unmindful that events beyond me had unfolded that would mark that day as one of some historical significance, both for the university and for the nation. That morning, the front page of the *New York Times* carried a story by William Beecher that disclosed the secret bombing of Cambodia. That afternoon, about the time I was addressing the issue of biochemical mechanisms of information processing in the brain, other students were confronting the symbol of their frustration inside Kansas Memorial Stadium as the chancellor prepared to review the ROTC units on parade. Barred from entrance to the stadium prior to the announced time, dissident students had gathered at the west gate. Emotions ran high as the gate remained closed. Finally, the students surged through it, forcibly entering the stadium to stop the review from taking place. Faced with the possibility of a violent confrontation, the chancellor called off the review.

President Nixon and his national security advisor, Henry Kissinger, were infuriated that their secret bombing of Cambodia, for which articles of impeachment against Nixon would be introduced in Congress five years later, had been revealed. Kissinger immediately called J. Edgar Hoover, director of the FBL, to set in motion wiretaps of numerous journalists and members of his own National Security Council staff. The first of these was authorized on May 12 and from them would grow more wiretaps, then secret and illegal operations leading to the break-ins and burglaries that became the substance of the Watergate scandal and its aftermath.

The interval between my oral defense and graduation was perfect for the noncerebral activity of moving Samson's lab. A brand-new building had been built to house the Department of Biochemistry and Physiology, as well as the Kansas Bureau of Child Research, which sponsored a number of projects at the University in Lawrence and the medical school in Kansas City. It had also become interested in setting up a neurochemistry research lab at the State Hospital and Training Center in the southeastern Kansas community of Parsons, where a few psychologists were already doing interesting work with behavior modification of mentally retarded children. One of these scientists was John Hollis, who was charged with setting up the lab. He came to hear me present my results at a research meeting of Samson's lab group in early March. He and I hit it off well from the start, and he recommended me for a postdoctoral position at Parsons a few days later. We worked out an arrangement that allowed me to split my time between Parsons, setting up the new lab, and Lawrence, where I could do work for which facilities at Parsons were still lacking. This avoided the necessity for me to move away from Carol, who needed another year of work in order to finish her doctoral degree.

I had hoped to avoid any association with the new building, but my postdoctoral position ensured that I would continue to do some lab work in Lawrence for at least a year, so playing a part in moving the old lab and organizing the new one was unavoidable. By this time, Dianna and I had acquired enough seniority to command the optimal locations for our respective desks, and we took the opportunity to position ourselves next to each other. Dianna and I consulted with each other continually, even though we were doing no experimental work together. We read each other's manuscripts, critiqued each other's experimental designs, offered each other advice and support on personal matters occasionally, and argued about politics frequently.

The day of my graduation from the University of Kansas dawned bright, cool, and clear. I walked back and forth across the campus that morning, enjoying the late-spring beauty of the hill where the citizens of Kansas had decided to raise a university out of the turmoil of the civil war a century earlier. It was fitting that Jerry and I should graduate together, having shared the same values and ambitions for four years. It would have been great to have Bob graduate with us, but he was no exception to the rule that few students in the humanities could complete their doctoral work in four years. Carol and Marilyn, a year behind us, would require another year or two to finish as well. This night, they kindly made a fuss over Jerry and me, taking our pictures in our caps and gowns with the three velvet stripes beyond price on the sleeve, the symbol of the highest academic degree attainable.

The processional began. We walked the long, winding path down the hill and past the lovely campanile dedicated to veterans of an earlier noble war, into the stadium that three weeks before had seen violent protest against a current conflict of less certain purpose. That night, there would be no protests. The ceremony was long, but uneventful, and the degrees were conferred by a little past ten. The years of academic striving had culminated in success; the work and worry of a thousand long nights and hundreds of cold, early dawns had come to fruition. The more difficult and elusive quest for professional success and recognition was about to begin.

[1]   Kandel, E. R. and W. A. Spencer. 1968. Cellular neurophysiological approaches in the study of learning. *Physiol Rev* 48: 65–134.

[2]   Schmitt, F. O. and F. Samson. 1969. Brain cell microenvironment. *NRP Bull* 7: 301–373. Schmitt and Samson were the general editors and dealt with additions and corrections from the contributors, so they got authorship credit. There is no question that the publication would never have happened without my intervention though. Also, at my suggestion, the word *Environment* in the title was replaced with *Microenvironment* to emphasize that the focus was on the immediate, local environment of neurons, glia, and the intercellular spaces among them. This was the start of the general use and meaning of the term *microenvironment* in the neurosciences.

[3]   Barraco, R. A., B. J. Klauenberg, and L. N. Irwin. 1978. Swim escape: a multicomponent, one-trial learning task. *Behav. Biol.* 22: 114–121. Rats were trained to escape from a tank of water by climbing up a rope. They learned to do this in essentially one trial, which made it superior to the Morris water maze published years later in which rodents had to learn the location of a submerged platform. Nonetheless, the Morris water maze became widely adopted, and my earlier version was essentially never heard from again.

15

Neurochemistry Becomes a Thing

The biochemistry of brain tissue has been studied since the earliest days when biochemistry developed as a specialized field of chemistry. One of the first scientists to analyze the chemical constituents of the brain was Johann Ludwig Wilhelm Thudichum, a German physician and scientist who spent the last half of the nineteenth century working in London. He extracted from over a thousand human and animal brains a variety of molecular constituents, including carbohydrates like galactose and lactic acids, and complex lipids related to gangliosides. For this, he is generally accorded the title of the father of neurochemistry. Books and book chapters on neurochemistry began to appear by the midtwentieth century, but the ultimate recognition that a singular and unified field has emerged from more general predecessors is the formation of a new scientific society. The first neurochemical society to be formed was in Japan in 1958, but no such society existed internationally until the mid-1960s.

Birth of the International Society for Neurochemistry

Meaning no disrespect to Thudicum, the father of neurochemistry for my purposes was Jordi Folch-Pi. A native of Barcelona, where he received an MD degree from the University of Barcelona in 1932, he carried out a number of important studies on metabolism before immigrating to the United States to become a research associate at the Rockefeller Institute just prior to the outbreak of the civil war in Spain in 1936. Folch-Pi's research at the Rockefeller revealed several new classes of lipids extractable from brain tissue and brought him sufficient acclaim to merit his appointment in 1944 as director of scientific research at the McLean Hospital, a psychiatric affiliate of the Massachusetts General Hospital. With the opening in 1946 of a new research facility designed by Folch-Pi, he was able to continue pursuit of his long-term objective of developing a simple, nondestructive method for extracting total lipids from brain.[1] Aided by Marjorie Lees, a graduate student receiving one of the first predoctoral fellowship from the US Public Health Service, Folch-Pi worked for years to gradually perfect a method for lipid extraction using chloroform and methanol with small amounts of water. When homogenized in this organic solvent mixture, most lipids dissolved into the lower of two liquid phases while hydrophylic "contaminants" remained in the upper phase. Those contaminants happened to include gangliosides, which brought the method of Folch-Pi and Lees[2] to my attention.

By the early 1960s, a number of biochemists around the world, but particularly in the United States and Western Europe, with some in Eastern Europe and the Soviet Union, were focusing on the biochemistry of nervous tissue and holding periodic symposia on the subject. Folch-Pi and some of his colleagues decided that a sufficiently critical mass had been reached to form an international society of neurochemists. A provisional organizing committee was set up at the instigation of Folch-Pi and Heinrich Waelsch, including (among others) Holger Hydén, Ernst Klenk, and a British neurochemist, Derek Richter. In 1965, Henry McIlwain and Richter drafted statutes for the new International Society for Neurochemistry (ISN), and articles of association for the Society were filed by solicitors in London, making the organization official. The first biennial meeting of the Society was held in Strasbourg in 1967. Folch-Pi was the Society's secretary for its first four years and became its chairman for two years from 1971 to

1973. Derek Richter was the Society's first treasurer and became its second chairman.[3]

From its earliest inception, the ISN was divided into two camps: one wanting to make membership highly exclusive based on a substantial record of research in neurochemistry; the other advocating a more liberal admissions policy to welcome investigators into membership earlier in their careers. Richter championed the more liberal view, but he was in the minority. The high bar for admission to membership in the ISN was a source of stress in the new Society, and it would come to affect me personally.

Birth of the American Society for Neurochemistry

About half of the original membership of the ISN worked in North America, and many of those members felt that a nationally based society, meeting annually in the Western Hemisphere, was needed to provide an adequate platform for the growing volume of research in neurochemistry taking place in North America alone. This, plus (*a*) the sense that the ISN was dominated by an elite group of researchers on the East Coast of the United States and (*b*) the feeling that clinical applications and behavioral studies were not valued highly enough by those in charge of the ISN, set the stage for a movement to create a society of neurochemists localized in North America (later expanded to include the entire Western Hemisphere). In 1968, two independent initiatives were launched. One was by a group of neurochemists at the Federation meetings in Atlantic City in April who authorized Martin Gál, an eminent biochemist in the Department of Neurology at the University of Iowa Medical School, to send out a solicitation of interest in the formation of a new American Neurochemical Society. Separately, Folch-Pi had decided to solicit an expression of interest in forming a neurochemical society specifically in North America.[4]

When Gál found out about Folch-Pi's initiative, he withdrew his own but retained the right to consider publication of a new journal. At the time, Pergamon Press was publishing the *Journal of Neurochemistry*

(JN) as the official organ of the ISN, but Pergamon was in a period of administrative turmoil and getting papers published in the JN was difficult. Besides the problem of publication, West Coast members were concerned about underrepresentation, and younger neurochemists were worried about the elitist standards for admission to membership already emerging in the ISN. Richard Lolley, one of Samson's earlier graduate students, wrote from California that "the West has enough neurochemistry to warrant a voice" and urged a genuine role for younger neurochemists.[5]

The first organizational meeting was held in November of 1968. An invitation was issued through a publication in *Science* the following February for all neurochemists interested in joining the new Society to get in touch with one of the founding organizers who could serve as membership sponsors. Several of us in Samson's lab responded and became founding members of the Society. At the November 1968 meeting of the organizers, however, it was decided that membership criteria in the ASN would be similar to those of the ISN; meaning, that a minimum of three publications in neurochemistry and status as an independent investigator would be required to qualify for ordinary membership. Those failing to meet the criteria were relegated to an inferior classification. By not specifying the value of different degrees of authorship (senior author, first author, etc.) or defining precisely what constituted being an independent investigator, the latitude for subjective judgments in determining who could and could not be an ordinary member was enormous. This would serve as a source of friction between young investigators like myself and the elders perceived to be autocratic elitists during the early years of both the ASN and ISN.

Another point of contention surfaced at the second meeting of the ASN organizing committee, hosted by Bernie Agranoff at the University of Michigan in June of 1969. Some of the organizers hoped to move away from the fifteen-minute oral presentations, then conventional at scientific meetings, in favor of more symposia, workshops, and small group discussions.[6] Folch-Pi, however, argued forcefully in favor of the traditional fifteen-minute platform presentation, and that carried the day. Thereby, the ASN passed up the opportunity to be truly innovative at its inception.

Francis LeBaron was elected by the ASN's organizing committee to be the first president of the Society, and he offered to host the first annual meeting on behalf of his home institution, the University of New Mexico in Albuquerque in March of 1970. Notwithstanding the inferior category of my membership, I would be eligible to present a paper and started making plans to do so as soon as the meeting was announced.

Meanwhile, Back in Kansas

As the new American Society for Neurochemistry was being born, I was assuming my duties as a Postdoctoral Fellow of the Kansas Bureau of Child Research. My primary task was to help set up a neurochemistry lab at the Parsons State Hospital and Training Center in Parsons, Kansas, about 125 miles (200 km) due south of Lawrence. The arrangement reached among Samson, John Hollis, and myself was that I would split my time between Parsons and Lawrence.

The time I spent in Lawrence during the first part of this period was devoted to completing a publishable experiment on ganglioside turnover in trained and untrained rats. Having succeeded in publishing one paper on the swim-escape task that I had devised, I was further encouraged to use the same behavioral manipulation for my latest version of the ganglioside turnover experiment. In essence, my results showed that the metabolic turnover of these compounds is influenced by the functional state of the brain. Though I wasn't able to demonstrate a correspondence specifically between ganglioside turnover and learning, it was the first demonstration that brain ganglioside metabolism is sensitive to non-injurious behavioral stimulation. After its publication in the fall of 1970 in the *Journal of Neurochemistry*, the paper was cited occasionally as an indication that gangliosides might play a role in the processing of information in the brain.

The most important thing I did in Parsons was to begin a long line of research on neurochemical aspects of the development of the retinotectal projection—the connection of nerve cells from the retina to the midbrain tectum—in the chick embryo. Convinced that the chemical basis of neurospecificity (the highly specific way in which one part of the nervous system hooks up with another) was central to understanding the degree of molecular specificity of information in the brain, and stimulated by the provocative ideas of Sam Barondes on the potential role of glycoproteins and gangliosides to neurospecificity, I decided to analyze the time course of the appearance of those molecules at different phases of development of retinotectal connections in the chick embryo.

On my way to Parsons the week that these thoughts on neurospecificity began to gel, I had given Marilyn a ride to the airport in Kansas City. She was flying to Houston for her first reunion with Jerry since he had moved there to assume his postdoctoral position at the Baylor College of Medicine. In Jerry's absence, Marilyn had not been socially inactive. My fear, in fact, had been that once they got out of sight of each other, they might get out of each other's mind. My concern was heightened as Marilyn, naturally enough, began seeing other men. It was only a matter of time until a woman as sharp, smart, and good looking as she was would attract any number of suitors into the void left by Jerry's departure. So I was pleased to play a role in dispatching her to Houston—though, given Jerry's history, not particularly encouraged to expect a significant outcome.

In that I was wrong. Perhaps Jerry, in solitude six hundred miles from Lawrence, had come to appreciate what we all had seen in Marilyn from the start. Or perhaps absence really had made their hearts grow fonder. Whatever the explanation, Marilyn came back from Houston engaged to marry Jerry, and for quite a few days, no happier person could be found in the state of Kansas.

[1]   Lees, M. B. and A. Pope. 2001. Jordi Folch-Pi, 1911–1979. *Biographical Memoirs* 79. Washington, DC: National Academy Press.

[2]   Folch-Pi, J, M. Lees and G. H. Sloane Stanley. 1957. A simple method for the isolation and purification of total lipids from animal tissues. *J. Biol. Chem.*

226: 497–509. This is one of the two most frequently cited references in all of biochemistry and one of the ten most-often cited publications in all of science.

[3] Bachelard, H. 1993. 25 Years of the International Society for Neurochemistry. *J. Neurochem* 61: S287, http://www.neurochemistry.org/Home.aspx.

[4] Tower, D. B. 1987. The American Society for Neurochemistry (ASN): Antecedents, founding, and early years. *J Neurochem* 87: 313–326. https://www.asneurochem.org/images/resources/History/ASN-history-1987-Tower.pdf

[5] Lolley, R. N. 1968. Letter dated May 14, 1968, from Richard N. Lolley to Jordi Folch-Pi, on the proposed ASN. [ASN Archives]

[6] Wolfgram F. J. 1984. Letter dated August 14, 1984, from Frederick J. Wolfgram to Donald B. Tower on planning for the first (1970) annual meeting of the ASN. [ASN Archives].

16

Interlude

Fred Samson (Fred, as I was now calling him) had become an institutional appurtenance of NRP by the summer of 1969 and therefore figured to play a role in the upcoming stellar event for which the NRP was best known—another of its Intensive Study Programs (ISPs). Patterned on a month-long conference at the University of Colorado in Boulder on biophysics that helped to crystallize and unify that discipline in 1958, the first ISP in the neurosciences had been held at Boulder in 1966. For four weeks, the world's most eminent experts in a great range of topics lectured and discussed their subject matter among themselves and specially selected junior colleagues of high promise (*eminence* and *promise* being qualities necessarily defined within the confines of NRP). The product of this scientific Rocky Mountain High was a hefty volume entitled *The Neurosciences: A Study Program*. Like its predecessor in the field of biophysics, this volume did much to define neuroscience as a distinct and emerging discipline, hybridizing anatomy, biochemistry, and physiology with a focus on the nervous system. Now three years later, a repeat performance was going to be attempted.

Fred was invited to attend as a member of Frank Schmitt's staff. My work on the "Brain Cell Microenvironment" bulletin the previous summer apparently earned me a similar designation. The letter of invitation was vague regarding my duties, and inasmuch as I was invited as a member of the staff rather than as an incipient neuroscientist of great promise, I did not take the invitation as a particularly great compliment. There could be no doubt, though, that participation in the meeting in any form represented an important—possibly a great—opportunity, so I gladly accepted.

Most of my career, I believe I have overrated more than underrated my professional success. My early opportunities at NRP, however, may be one case in which I underrated my value as seen by others. Fred always insisted that NRP viewed my contribution as valuable, independent of my relationship with him. Since I came to believe later that Frank Schmitt was shrewder and more purposive than he at first appeared to me, I am prepared to believe that I was asked to be a part of the ISP in my own right at Schmitt's direction in 1969.

On July 2, barely a month after I had submitted my paper on subcellular changes in protein and sialic acid to *Brain Research*, the editor wrote back that he would be pleased to publish it if I would make just a few minor changes. For the first time in my experience, a journal had accepted a paper immediately and almost without qualification. I was quite thrilled, of course, and derived from that good fortune a false sense of confidence about my impending impact on the field of learning and memory.

"When Ships to Sail Among the Stars Have Been Built"

As my professional fortunes seemed to be rising, the spirit of the nation and the history of the world were counting down to a momentous event. The race to the moon was no longer a contest between the United States and the Soviet Union, but it remained a race against time. President Kennedy had committed the United States to a manned lunar landing by the end of the decade. With seven months to go, the nation launched its attempt to live up to this legacy. Three American astronauts—Edwin Aldrin, Neil Armstrong, and Michael Collins—were hurled aloft from pad 39-A at Cape Canaveral the morning of July 16 on a voyage of epic and unprecedented proportions.

While those three were en route to the moon, I was en route to Denver, having been summoned to Boulder a few days early by Fred, who said that Frank Schmitt wanted his team there and in place early. I settled into my assigned room in one of the dormitory towers on the University of Colorado campus and gathered with others on Sunday afternoon, July 20, to watch live coverage of the moon landing in the top-floor television lounge. I was there early to get a front seat for the coming show. At just after two o'clock Colorado time, the lunar module bearing Armstrong and Aldrin separated from the command module and began its three-hundred-mile arc toward the surface of the moon. A quick view out the dormitory window disclosed a deserted campus and streets beyond with hardly a moving car in sight. The world had come to a standstill. At 2:17 p.m., Armstrong radioed that the lunar module had landed intact and upright in the moon's Sea of Tranquility. Walter Cronkite was speechless. Most of the spectators in the lounge with me applauded, but I was too awed to move. The moment was profound beyond celebrating.

At 8:56 p.m., Neil Armstrong placed the first human footprint on the moon. Aldrin followed him onto the surface nineteen minutes later. They romped their way around the module for the next two hours, setting up experiments, collecting rocks, giving guided tours and geological lectures to the television cameras that NASA had wisely included, sharing with billions of viewers on Earth the thrill of their endeavor. It was after 11:00 p.m. when the astronauts returned to their module for a few hours of rest before initiating their return to Earth. I went to bed well after that, grateful to have seen one of the most ancient of human dreams come true. "When ships to sail the void between the stars have been built," wrote Johannes Keppler centuries ago, "there will step forth men to sail these ships." And so there were.

It was a sign of the times that there would be those who found fault with this trip to the moon. The cost could better be spent on the desperate social needs of this planet, they argued, failing to see the voyage as an achievement of the collective human spirit that, for a few brief hours, brought a sense of unity to the world that never had existed before. For a true and pervasive empathy to develop among all the people of the planet, we have to have a feeling that we have a stake in one another's lives. Religion and introspection bring this revelation to some, but only a worldwide event to which the whole world is witness brings it into the collective experience of people in all lands and walks of life. The voyages to the moon, especially this first one, provided just such a unifying experience and provided it to a greater degree than anything before or since.

Getting to Know the Best and the Brightest

Monday, July 21, 1969, astronauts Armstrong and Aldrin successfully lifted off the surface of the moon and rendezvoused with their mother ship in lunar orbit, rejoining Michael Collins. Soon, all three had blasted free of lunar gravity and were headed for home. With the lunar mission an apparent success, Schmitt and the other managers of the ISP breathed a deep sigh of relief that no catastrophe had occurred to dampen the start of the meeting. On a high note of optimism, then, an intensive three weeks of presentations and discussions by supposedly the best and the brightest in the neurosciences of 1969 got underway.

I never knew how much Schmitt really heard or understood. Certainly he understood more than he appeared to, but it seemed to be his technique to act as disorganized and disoriented on the issue of the moment as possible. It was as though by so doing, he was better able to bring out the input from those of us around him that would trigger his genius for ultimately expounding on the subject with flair and zeal. Getting to that ultimate product, though, required that his staff suffer through interminable meetings that can be characterized as chaotic and confused at best. The first of those meetings ensued within an hour of my arrival in Boulder.

Suffering with me on the staff were Howard Wang and Barry Smith. Howard was a recent graduate student from Ross Adey's lab. He impressed me greatly from the beginning, in part because his research was consonant with my growing ideas about distributed, nonsynaptic localization of information storage in the brain. Barry Smith, on the other hand, did not impress me in the beginning. He was Schmitt's last graduate student and seldom seemed to have much of anything to say independent of Schmitt's own ramblings. As it turned out, there have been few people that I more thoroughly misjudged in the beginning than Barry, as events a few years later would reveal.

On the whole, the quality of the meeting was indeed high, and it provided me with a memorable experience at a critical point in my professional development. Not that it wasn't disappointing in some respects. There was a frustrating failure to relate the molecular biology and biophysics expounded with such elegance to the equally impressive psychology and developmental biology that was presented. There wasn't even an attempt to define the extent of the gap. The one scientist who might have tried was Holger Hydén, the first and boldest searcher for memory molecules, but he was unable to attend. Of the many individuals and specific events that did make the meeting memorable, I will briefly relate only a few that bear on my central story.

Paul Weiss, the towering figure in developmental neurobiology for half a century, underlined as a central theme of the field the notion that nerves and muscles have or behave as though they have individual identities. This man, who appeared to me as a genteel elder statesman in my conversation with him on a lawn in Weston, Massachusetts the previous summer, was provoked into a more feisty and argumentative demeanor by Marcus Jacobson, a brash and abrasive embryologist, formerly from South Africa, then at Johns Hopkins. Jacobson had carried out a very elegant set of experiments demonstrating that nerves become specified to a highly unique degree by the tissues they connect to, meaning that a particular sensory neuron will form functional connections with a particular patch of skin, but not other patches. His arguments with Weiss, who had earlier demonstrated what appeared to the outsider to be essentially the same phenomenon, revolved around rather technical details; their clashes, however, were vigorous and seemingly symbolic of a deeper chasm—a generational gap or perhaps the conflict of two stubborn and aggressive personalities. Whatever their debate really meant, it was an intellectual clash of the first order. I rather enjoyed it, admiring Marcus Jacobson, the brash young upstart, as much as Paul Weiss, the venerable master. Little did I suspect that, years later, I would find myself the object of Jacobson's ire.

An immunologist from Rockefeller, Gerald Edelman, presented a brilliant overview of the molecular mechanism of the immune response. Whether it was the beauty of the scientific argument or the lucidity of its presentation by Edelman that impressed me most, I can't be sure; but one way or another, I left the auditorium that day predicting, along with many others, a Nobel Prize for the speaker. In 1972, our predictions came true. In 1977, from virtually the same podium, this brilliant scientist would present an eloquent theory of higher brain function with the same force and apparent insight that he had brought to the discovery of the molecular structure of antibody molecules.

Theodore Bullock, coauthor of an encyclopedic and monumental opus on the nervous systems of invertebrates and alone among NRP Associates in his appreciation of the historical (evolutionary) as well as functional determinants of neuroanatomy and neurophysiology, chaired a symposium on neural systems. In essence, his symposium argued that there are many ways for neural subsystems to organize and transmit information; exclusive focus on one site (say, the synapse) or one mechanism (say, action potentials) is too confining. "Our vision is limited by our success," he said. "We open up rich veins and dig for ore with the tunnel vision of a miner." At NRP's Stated Meeting of Associates the previous summer, Bullock had given what I considered an uninformed and mundane reply to one of my questions, so I had brought an unfavorable impression of him to Boulder. Now he was saying what I was coming to believe wholeheartedly, and saying it with eloquence. Another mistaken first impression. As I listened more carefully to him then, and in the future, I came to realize that if anyone was going to bridge the gap between the molecular and the behavioral domains of the brain, it would more likely be him than Hydén or any of the chemists.

Floyd Bloom, a neuropharmacologist at St. Elizabeth's hospital in Washington, was a rising star with NRP, but not nearly so eminent as Edelman or Bullock. He earned my grudging respect at the ISP despite his focus on the synapse because of the thorough way he analyzed all the possibilities, both structural and chemical, for synaptic involvement in learning. Unlike most scientists at the time who were much more willing to publish theories than experiments about the role of synaptic junctions in learning, Bloom at least had carried out a number of behavioral manipulations in search of some correlation between behavioral stimulation and the number or function of synaptic junctions. His insight that our main impediment to "a molecular explanation of neural plasticity is that we do not yet know where to look nor what to look for" was praiseworthy in my view. What I most remember him for, though, was a remark near the end of his talk. "Since this is 1969, we decided to try cyclic AMP," he said, thereby implying that experiments that would turn out to be among the most productive in his career had originated simply as the thing to do because everybody was doing it at the time. I doubt that his motive in using cyclic AMP was really that aimless, but it marked him early in my mind as an opportunist with an uncanny instinct for the fruitful tack to take in research. The coming years would see this repeatedly confirmed, as he rose eventually to superstar status at NRP.

Robert Galambos chaired the sessions on determinants of neural and behavioral plasticity, the topic closest to my own interests. Because of my involvement as a staff member in Schmitt's concurrent program on molecular neurobiology, I was not able to work with Galambos as I would have preferred. As the meetings were nearing their end, Galambos got wind of the fact that I had worked with Ungar and had strong ideas about research on learning and memory. He told me that he wished he had known about me sooner, implying that he could have used my help. *What an irony,* I thought. One of the first books I had read, still as a teenager during the hot nights on my cot in Kermit, after long days of chasing lizards in the sun in the summer of 1962, was the little book by Robert Galambos, *Nerves, Brain, and Muscle.* When I had begun to look into graduate schools, he was one of the first neuroscientists I had written to, then at Yale, and one of the first to encourage me. Though his prediction that I would probably have no trouble getting into Yale did not come true, I didn't hold it against him. I held him in respect and esteem, but in the abstract as I had never met him. Now seven years after the summer in Kermit, he was telling me he wished he had known about me sooner. It almost made up for not getting into Yale.

Sam Barondes, I met for the first time at the start of the meeting in the television lounge where we all watched the lunar landing, as he was staying in a room near mine. Thus we had a chance to get to know each other informally before his two formal presentations. His low-key, very relaxed demeanor belied my stereotypical expectation based on his New York City upbringing. I gathered that one (though certainly not the only) secret to his success was his ability to put others at ease. His two talks in two separate symposia at the ISP reflected the crossroads at which his career had arrived. The first presentation, "Multiple Steps in the Biology of Memory," reviewed the work that he, McGaugh, Agranoff, and others had done to establish the fact that long-term memory formation required the production of protein in brain cells and that it was probably a multistage process rather than a singular, instantaneous event. "The difficulty of drawing conclusions about such biological processes from behavioral studies is recognized and bemoaned," he concluded, in simple but elegant testimony to the frustration of those of us who had labored as yet unsuccessfully to infer from an animal's behavior what the molecules in its brain were doing.

Barondes's second presentation was actually of greater immediate interest to me because it dealt with the proposition that glycoproteins and gangliosides might mediate the specificity of neuronal connections and their function. In a purely speculative model, he showed how gradients of carbohydrate chain length and number of charges such as those carried by sialic acid residues could generate the type of positional information that would allow specific cells in the retina to hook up with specific cells in the tectum. The ideas in his talk were not original with him, but he brought together a number of different notions and speculations with clarity, showing why the macromolecules of the brain with carbohydrate residues attached to them deserved the serious consideration of brain researchers and pointing to how this could be done experimentally. It turned out, of course, to be a lot easier to say than to do, but he pointed the way. The publication based on this talk became one of the most frequently cited references on the possible importance of glycoproteins and gangliosides in the neurochemical literature.

Meeting all these scientists for the first time was obviously exciting and beneficial, but none did I prize more than my meeting with Ross Adey. It was ironic, I suppose, that Adey, who was not a chemist to speak of—his neurophysiological methodologies in fact were quite beyond my understanding—should have come to influence my thinking as much as he had. But from the first of his series of papers on learning in the early '60s that I had read, I felt instinctively that his view of how the brain operates was the most cogent and the closest to reality of that of anyone writing on the subject. While I didn't understand the electrophysiological details of his research, I comprehended the broad outlines well enough to know that he, too, was visualizing information as dispersed patterns of fluid bioelectrical states rather than as a sequence of punctate switches, and a pattern that may well arise from micropatches of membrane whose bioelectrical excitability, hence function, is locally controlled by surface molecules such as glycoproteins and gangliosides. In our first meeting, as on every occasion subsequently, he seemed genuinely pleased that I had been able to draw insight and inspiration from his work. He was a pioneer in the field of electroencephalography, a major investigator for NASA in its early studies on the physiology of space flight, an advisor to presidents. I marveled that from our first meeting, he always cared what I thought.

Howard Wang and Ross Adey had recently shown that calcium injections into the cerebrum affected the impedance, hence bioelectrical conductivity of the extracellular space of the brain. In one of the afternoon sessions open to discussion from the floor, Howard had talked about their research. This offered the perfect opportunity for me to describe my swim-escape experiment with its sialic acid changes in the dendritic, but not synaptic fractions. I did so in about a two-minute presentation, at the end of which I assumed there would be appreciative comments concerning the experimental support that I seemed to be giving to the Adey-Wang suggestion of the importance of glycoconjugates to information processing in the brain. In fact, some explicit congratulations would not have been out of order. Instead, there was total disinterested, and apparently uncomprehending, silence.

Howard, of course, appreciated the relevance of my experiment to the issue of information representation in the brain. We had talked about it earlier, standing in front of a gorgeous artistic rendering of the dense dendritic forest of the cerebrum, telling ourselves how obvious it seemed that it was there, through the branches and leaves of the dendritic arbors, that information danced and dissolved in the process of moving a finger, informing the consciousness of a sight or sound or, in some cases, leaving a tell-tale trace that a subsequent perception or behavior would betray as memory. That my crude chemical analysis of dendritic and synaptic cell fragments from brains of rats thrown into a tank of water transcended the gap from molecule to memory, I was not asserting. But that the flicker of a relationship between my experiment and the lofty generalizations to which the ISP presumably aspired could be divined in the mind of someone in the audience, I did feel was a reasonable expectation. It apparently was not. It was, instead, the first, but regrettably far from the last, indication I would get that the significance of my work was much greater in my own eyes than in the eyes of my peers.

Questions Unasked

The ISP wrapped up at the end of the first week in August, and Carol and I took our first train ride together—a trip to California to visit friends. California appealed to my taste for the different and distinctive, but aggravated my discomfort with extremes. Every exaggerated aspect of society in the late 1960s seemed to be present in abundance in California. The tension was palpable, and only half a summer of living in laid-back Santa Barbara many years later enabled me to finally overcome that feeling of tension whenever I entered the state. So it was not with a great deal of regret that we returned to Kansas, where my postdoctoral duties and time, split between Parsons and Lawrence, resumed.

In mid-November I attended a work session on macromolecules in synaptic function at NRP—again invited as a participant but not a speaker for no obvious reason other than that Samson was invited. The two-day conference was a disappointment: no unifying concepts emerged nor were any very challenging questions raised. What really disturbed me was the failure of all those accomplished neuroscientists to ask more important questions: Can macromolecules at the synapse be changed by the level of synaptic activity, and if so, are the changes reversible? Do synaptic connections have to be plastic for the brain as a whole to be plastic? Are there physical or structural alterations undetectable by chemical means that could influence the function of a synapse?

So why didn't I ask these questions myself if I thought they were so important? For two days I sat there, thinking surely someone would bring up a query or draw a connection that extended slightly beyond the data of the speaker at the rostrum. Save for some spirited debate over exocytosis, which began to take on the tone of the ontological argument, there was nothing offered by anyone that justified the expense of bringing all these people together, other than the benefit to the participants themselves of getting together to share their latest information before the rest of the scientific community got access to it.

But why didn't I speak up? This was the question I tried to confront at thirty thousand feet over western Massachusetts, a Bloody Mary in hand to dull my impulse to rationalize. The fact of the matter was that I had felt just a little too intimidated to speak up in that august company—not because of anything they did or said; it was all in my mind and all my fault. I had, in fact, finally spoken up at the postmortem of the meeting held between NRP staff members (of whom I was apparently considered one) and the meeting's organizers the morning after the last session, and they all seemed to think my questions were good ones that should have been addressed. I was glad that I finally spoke up, but all the sorrier that I hadn't done it sooner. I had missed a prime opportunity at a critical time in my career to speak out against the lack of synthetic and integrative thinking in neuroscience that I criticized so often among friends. I would later come to realize that the loss for neuroscience was not so great nor the damage to my career so devastating, but I still wish I had been just a little more aggressive. I wish, like all the shy people in the world, that I weren't so shy.

Energized in Albuquerque

Since Fred and several of his graduate students, including me, were founding members of the ASN and thus eligible to present papers at the society's inaugural meeting, we anticipated the trip to Albuquerque in March 1970 with considerable excitement. For me, it provided the opportunity to follow up on the ganglioside turnover experiment that I had reported in Atlantic City the previous year, by presenting my results to a more appropriate and presumably sympathetic audience. For Dianna, it meant her first presentation of a scientific paper at a national meeting. For Hugo Fernandez, it was the first opportunity to report on work he had done outside of Frank Schmitt's lab, where he had done his doctoral research. For Stan Twomey, it was a dual opportunity to shine in his inimitable way, as he presented two papers at the meeting. And for Samson, it was an opportunity to portray his lab as a leader in the newly coalesced field of neurochemistry. At that first meeting of the ASN, he coauthored four presentations, more than any other member of the society.

Three of the four papers presented from Samson's lab dealt with microtubular protein (tubulin), the substance thought by then to be responsible for the transport of material from its point of manufacture near the nucleus, down the long axonal or dendritic extensions to the synaptic endings of the cell. Only my paper with Samson on gangliosides had nothing to do with tubulin. The papers on tubulin by Hugo, Stan, and Dianna were all scheduled for presentation the same afternoon as mine in a different concurrent session. Thus Fred had to choose between attending my presentation or the presentations of his three coauthors presenting the tubulin research. There was never a question in my mind that he would attend the tubulin sessions rather than mine since his own interest and involvement were greater with the tubulin work than with gangliosides. Furthermore, as he explained apologetically, I was his senior and most experienced student and needed him less. I fully understood and accepted his decision, but allowed myself to pout about it a little, nonetheless.

My presentation was adequate, but not as good as I would have liked. My paper was the tenth of ten scheduled presentations, the second afternoon of the meeting. By four thirty in the afternoon of any scientific meeting, most scientists are bleary-eyed and restive, so attendance falls off sharply as the cocktail hour approaches. There were two or three neurochemists in attendance whom I knew had an interest in behavioral neurochemistry, and only a handful of others. Among the former in the audience was Georges Ungar, who complimented me on my presentation, but confessed that he couldn't quite see how gangliosides would be as interesting or important as peptides.

Ungar had come to Albuquerque in higher spirits than I had seen him in a long time. Three years earlier, he had seized upon a simple passive avoidance task in which rats were trained to avoid their natural tendency to escape into dark compartments by giving them mild electric shocks when they did so.[1] In Ungar's hands, this had turned out to be a very robust learning task that could reliably be transferred by injecting peptide solutions from donor brains into naive mice.[2] A couple of months earlier, he had finally isolated and chemically characterized the peptide that he believed to be responsible for the transfer of dark-avoidance in mice. In a scheduled presentation the third afternoon of the meeting, he reported that the material was a peptide consisting of fourteen amino acids (the actual number turned out to be thirteen), and he reported what the amino acid composition was. Privately, Alberte told me that Georges was extremely pleased because he really believed that his career was going to come to an end on a triumphant note with discovery of the first molecule specifically related to a mechanism for encoding memory in the brain.

On my return to Lawrence, I detoured through Lubbock for an eighteen-hour stopover. After dinner and a good visit with Kenneth Davis, my former English professor and only friend still living in Lubbock, I took a walking tour of the campus and surrounding area to rekindle my memory of the good times and the bad that I had lived through there in the earlier years of the decade. The classroom where I first met Penny May and encountered Don Tinkle was unlocked, so I walked in, sat down in the center of the front row as I had done the first day of classes in 1961, and thought about them both and the times we shared. I walked further, past the Student Union where my attempt to organize a symposium on campus political parties had raised the ire of the administration in my freshman year. I passed the building where I ran down three flights of stairs and all the way home after scoring 96 on a calculus final in a course I was barely passing. Then I strolled past the steps where a friend and I had sat through the night, idly scribbling chalk marks that remained on the steps for months as a reminder of the grave conversation we had shared.

After midnight in a freezing drizzle, I walked to the corner of Main Street and Avenue X, then looked back across a city block that had been leveled for the construction of a parking lot and a gaudy new motor hotel. Houses and apartments once stood there where Kay and I and a number of our friends had lived. I walked to the spot where the basement would have been, where I spent so many late nights and early mornings learning organic chemistry and comparative anatomy, writing my first scientific paper, struggling with the problems of the present, and thinking about the possibilities of the future. Where once I had done all those things was now nothing but mud and weeds, as though it had never really happened, as though we had never existed. Rain whipped into my face, and the late winter wind chilled the back of my neck. For a moment, it was one of the most melancholy moments of my adulthood. I walked back to my hotel room for a few hours of sleep before catching the plane back to Kansas the next morning. Carol was a candidate for a research award, and I wanted to be home in time for the ceremony.

Moving On Again

I returned from Albuquerque and Lubbock to a tense community in Lawrence, simmering with discontent over racial tension at the high school and events in Southeast Asia. Publicity from the unrest of the previous year, such as disruption of the ROTC review in May, turned the University of Kansas into a national magnet for students and semi-students with a radical bent. The vast majority of us who viewed ourselves as moderates strained to hold the center, anguished over the nation's inability to come to grips fully with the race issue or get out of the Southeast Asian quagmire, but not convinced that violence would achieve either of those goals.

As the 1960s ended, the casualties continued to mount in Vietnam and public support for the war continued to fall, but the American people, ever anxious to give a new president some operating room and the benefit of the doubt in the beginning, limited their objections at first to large but peaceful protests. But then, Nixon decided to invade Cambodia with ground troops in support of a coup in that country against the neutralist Prince Norodom Sihanouk. Four days after the Cambodian incursion, four students were shot to death by frightened national guardsmen on the campus of Kent State University. And in Southeast Asia, the Cambodian incursion set in motion events of terrible consequence, generating a communist insurrection that led years later to a genocide in Cambodia of horrible dimensions.

While all this was happening, I was trying to get something accomplished in my year of postdoctoral study. The dominant new figure in my life, both personally and professionally, was John Hollis, the research psychologist at Parsons directly responsible for recruiting me. John was imaginative, intelligent, and blunt. He had little patience for rigid and unimaginative colleagues of normal intelligence (and took no pains to disguise the fact), but was perceptively tolerant of the limited but special abilities of the mentally retarded children with whom he worked. By a brilliant application of operant techniques learned as a student of Harry Harlow at the University of Wisconsin, he was one of the first psychologists to achieve any success at modifying the behavior of autistic children. He was a central figure in making the Parsons State Hospital and Training Center in Kansas nationally known for its progressive research in mental retardation in the 1960s.

I was well aware of the meaning of operant behavior and of its importance in psychology. But it was John Hollis who showed me the power of this form of behavioral modification and the technical details of its manipulation, which proved very valuable when, years later, a colleague and I conducted a set of experiments on the neurochemical correlates of operant training in pigeons.

John had wanted to move in a more biological direction and had recruited me for that purpose. We hit it off well and got a lot done in my short year of commuting between Lawrence and Parsons. I was able finally to complete a publishable ganglioside turnover experiment and initiate research on chemical changes in the retina and brain of developing chick embryos. But in reality, my postdoctoral appointment of just one year was a stopgap measure to keep me in Kansas while Carol worked on completing her own doctoral research. In truth, I needed more training and should have sought it. But the war had reinforced my desire to move out on my own as soon as I could, to start giving back to students and getting to a position of greater influence. As a student myself, I had been a beneficiary long enough—from Tinkle, Ungar, and Samson; from my local draft board; from the state of Kansas; and from NASA and the National Science Foundation. The war exaggerated my feeling that I had incurred a debt, the repayment of which I could stand to put off no longer.

In response to one of many inquiries concerning faculty openings at colleges around the country, the chairman of the Department of Biology and Pharmacology at the College of Pharmaceutical Sciences, Columbia University, called me on December 17, 1970, inviting me to New York for an interview. I gladly accepted, not daring to hope that I would end up living so soon in the heart of the most exotic place I had ever been to.

When within just a few days, the offer of a position as assistant professor of biology officially came through, Carol and I were ecstatic. A couple of other possibilities were pending, but we had decided almost from the start and with little discussion that if we got a chance to go to New York, we would jump at it. We were frankly awed by the city, and the prospect of moving from the esoteric atmosphere of a college campus into the churning chaos of the ultimate urban setting was an experience that we welcomed almost as a duty, given the temper of the times.

My impending move to New York posed two problems, however. First, I would have to go without Carol. She wasn't quite finished with the research for her doctoral dissertation. Dick Himes, her advisor, had agreed that she could write her dissertation in New York, but he wasn't about to let her get out of Lawrence without carrying out that "last, definitive" experiment. Our early and wildly overoptimistic estimate was that she would have to stay in Lawrence for perhaps a month after I left.

The second problem posed by my imminent departure was the need to bring some closure to the projects with John Hollis, barely after they had begun and far too soon to see John's vision of a viable neurochemistry lab at Parsons realized. If I have a regret about my last year in Kansas, it is that I wasn't able to help much in bringing John's vision to reality. Our relationship and scientific communication, fortunately, would persist nonetheless for several years.

The night of August 26, 1970, our fellow graduate students held a farewell dinner party for us. Marilyn and Jerry were married by then and living in Houston. Bob and Muriel had left Lawrence a month earlier for Bob to accept a faculty position in philosophy at the University of Evansville. I spent most of the night following the dinner loading books and other possessions into a rented U-Haul van. The next day, I said goodbye to Carol; then on my way out of Lawrence, I stopped by the lab to say goodbye to Fred Samson, the man who had served as a mentor in absentia during the first part of my graduate studies. He had returned to Kansas after I had become set in my ways, but had given me the support and latitude to follow my own vision and make my own mistakes about what was important in science. Dianna was not around, so I just left her a note thanking her for our friendship and predicting great accomplishments for her future, which time would see come true.

That night, I drove east to St. Louis, leaving the western foothills of the Ozarks where I had lived and learned so much in five short years. I stayed overnight with Art and Marlyce Friesen, renewing all too briefly that vitally important friendship from my first two years in Lawrence.

The second night I stopped outside of Akron, and the third night I spent in Western New Jersey, poised on the outskirts of New York City. The morning of August 30, 1970, I crossed the George Washington Bridge into Manhattan at 8:02. At 8:05, I was stopped by a policeman for driving a commercial vehicle on the West Side Highway.

1   Gay, R. and A. Raphelson. 1967. "Transfer of learning" by injection of brain RNA: A replication. *Psychon Sci* 8: 369–370.

2   Ungar, G., L. Galvan, and R. H. Clark. 1968. Chemical transfer of learned fear. *Nature* 217: 1259–61.

EARLY SEVENTIES
(NEW YORK)

17

Ice Buckets on Broadway

From the first few minutes of my life in Manhattan, I lived in a torrent of overstimulation. There were moments of quiet reflection, of course, ways in which I could get away from the churning mainstream of the vibrant life of the city. Like all New Yorkers, I learned to seek and defend my private places, even in a crowd. It didn't take me long to understand that the cordiality so reflexive in a southwesterner like myself is out of place in the pedestrian and subway-packed masses of the eastern seaboard—a stratagem necessary for survival in an environment bursting with people and overflowing with sights, sounds, odors, and impressions that bombard the senses without letup unless a way is found to filter them down to a stream of manageable intensity.

But that was also part of the beauty of life in New York for me: the knowledge that the stimulation was there to be let in at my slightest inclination. The sensory bombardment was pervasive, and I admitted every bit I could handle. The odor of roasting coffee that wafted in every morning from the processing plants across the Hudson River in New Jersey was one of my earliest, most delicious impressions of Manhattan. I was fortunate to find a two-room apartment on the fifth floor of a residential hotel on West Seventieth Street, just off Broadway for $250 a month in the fall of 1970. From my window, I could look down on the West Seventy-Second Street express subway stop and watch the river of automobiles flow like lava around it, their horns never silent, their passengers harried and frequently irate at their imprisonment in the mobile cages that they felt compelled to drive to and from their piece of privacy in the suburbs. I really enjoyed and felt a little guilty for the fact that some of them were spending hours in their cars every day while my work was a five-minute walk away.

In fact, liberation from dependence on a car was one of my favorite things about living in New York. Just about anything I could want, and much more than I ever imagined, could be found within easy walking distance of our apartment. There were five movie theaters within four blocks. Lincoln Center was five minutes to the south on foot. We walked in off the street for concert tickets at the Philharmonic, saw Bernstein's *Mass* on its opening night in New York, and enjoyed professional ballet for the first time, strolling back home after each of them in just a matter of minutes.

While the attractions of high culture were seemingly endless, it was the local culture of the street as much as the talent of the concert hall that made life in the city such an adventure. For a kaleidoscope of spontaneous entertainment, nothing beat a stroll through Central Park on the first warm, sunny Sunday of spring. The promenade of mothers, fathers, and toddlers along the West Side Parkway provided a hundred scenes a minute of fascinating family street theater, especially after we joined the cast with our own toddler our second year there. At the time we lived in New York, "Rose in Spanish Harlem" was a popular song, and I loved the way the music of that song and others like it poured out the open tenement windows of Spanish Harlem on warm Saturday mornings as I walked through the upper West Side on my way to use the library at the Columbia College of Physicians and Surgeons.

Naturally, the street life of New York had its seamier side, and I never got used to some of it. I don't know what appalled me more: my first sight of beggars lying on the sidewalk or the indifference that everyone showed toward them. At first, I would give them money and occasionally try to talk to them. But like everyone else, I soon became overwhelmed by their numbers (not excessive, in comparison to some parts of the world, but many more than I had seen any place I had ever lived). I never learned to be indifferent, but I got to where I could usually ignore them at the cost of a minute or two of guilt each time I passed them by. At least I was able to rationalize that in teaching at a pharmacy school, I was helping address the educational and social needs of the drug-riddled society around me. How a stockbroker handled the issue, I couldn't imagine.

Examples of opulence were too easy to see in the streetscape as well. The contrast between the mink coats and limousines that bristled past the beggars on the sidewalk was stark and painful to me. The disparity between poverty and prosperity seemed to be accentuated by the building boom underway in New York at the time. With a war raging in Southeast Asia, a country still wounded from race riots and political assassinations, and a city where poverty and prosperity coexisted without self-consciousness, it occurred to me in my euphoria over finding myself a resident of New York City at the start of the '70s that our society was faced with problems of considerable severity. I wanted badly to contribute to resolving those problems. They became the backdrop of my two greatest passions: to teach biology and to learn the molecular mechanisms of memory.

Monday Night Football

When the time came for me to move to New York, Carol was not ready to go because she had one or two last experiments to complete for her dissertation under the direction of Richard Himes. For Himes (as, in fairness, for many doctoral advisers) the term *one last experiment* was a euphemism for one last year of lab work. That Carol managed to get away from Lawrence within eight months of our separation has struck us in retrospect as a considerable accomplishment. In September of 1970 though, we anticipated that she would be along to New York in a month or two. So I had driven east without her and didn't see her again until Thanksgiving.

Alone in New York, consumed by my new job as an assistant professor of biology, and overwhelmed by the intensity of the urban setting, I settled into a bachelor's rhythm of hard work and little play. I was up early and off to my office for conferences, classes, and labs all day; then home in the evening as late as I felt like for a dinner of my own Spartan cuisine and additional hours of preparation for the following day. Monday nights were a special treat because the American Broadcasting Company, in a stroke of programming genius, inaugurated *Monday Night Football* that September. I would be home by six on Monday nights to start my preparation for Tuesday's lecture and would work like a demon till nine. The football game would start as I fixed my supper, which, by the end of the second quarter, I would have consumed. Howard Cosell's halftime highlights were a real treat, but if the lecture were not finished, I would forego that pleasure in order to polish off my notes for the next day. Then in the third quarter, I would return to the couch feeling virtuous and relaxed enough, usually, to fall asleep before the game ended and the nightly news of muggings and mayhem in the city outside my doors came on.

I worked very, very hard that first autumn in New York, as I had to design and teach an entire course with lab in general biology for the first time. For years, it had been my dream to do just this, and I loved doing what I had hoped for so long to do. The students were interested and engaged. Though not used to the lengthy take-home essays that I gave instead of the objective in-class exams they had grown to expect in science courses, they threw themselves into the assignments with energy, exhausting themselves in the effort and me in the grading that followed. I introduced small animal surgery, which they loved, and talked about things like ecology and evolution, which, as future pharmacists, some of them came to accept only grudgingly. By the end of my first course at the Columbia College of Pharmaceutical Sciences, my student evaluations were the highest of any member of the faculty.

A Significant Reconnection

Through the fall, Carol worked desperately to finish the work on her dissertation, writing every month to tell me she thought she would be with me in one month more. It turned out that she didn't get to New York till Thanksgiving and, even then, had to return to Lawrence for another three months after Christmas. While I was content enough to manage on my own, and my heavy workload was an effective distraction from loneliness, I nonetheless missed her very much. She had loved New York before I had ever seen it, and I knew she would relish it as much as I once she arrived. I couldn't wait to share this marvelous new experience with her.

Not filling the void, but easing the pain of Carol's absence for me was my close friend, Penny, from our days together at Texas Tech and in Houston. Fate had brought us to New York City at the same time. She had completed her graduate work at Tulane and procured a postdoctoral appointment with Dorothy Bliss, a noted invertebrate endocrinologist, at the American Museum of Natural History about twenty blocks north of where I worked. Having lived for several years already in the multicultural mélange of New Orleans and born with a cosmopolitan outlook and air anyway, Penny fell into the lifestyle of New York City with ease and helped explain it all for me. After the years of separation and all that we had been through, we were together in the same city again—incredibly far in so many ways from the deserts of West Texas.

That first semester I was inundated with preparations and grading for my course, so getting research going was slow. It wasn't until Thanksgiving, the day before Carol's arrival, when I initiated my first lab work in New York, and it wouldn't be till the spring semester that most of the student projects got underway. Working on Thanksgiving Day, in the peace and quiet of the lab with no lectures to prepare and no students to see, I was able for the first time to take a deep breath and assess my research situation. On the positive side, the lab had a fantastic view. I could look out on Columbus Circle and the General Motors building beyond it to the south and, on a clear day, see down to the Empire State Building. At night, the scene was magnificent— each building a shining chandelier thrusting into the nighttime sky. It had only been six years since I had stared in awe at the skyscrapers of Chicago. Now to be working in a place where any day I could look out the window and see a dozen or more among the world's tallest buildings amazed me every time I thought of it. Often I would get up from my desk and walk into my lab, just to look at the view and to ask myself if I were really there.

On the negative side was almost everything *but* the view: The space was small and cramped. There were no major pieces of up-to-date equipment. The building was old (this was the second oldest college of pharmacy in the country, and the building was built before the turn of the century) and structurally inadequate as a teaching facility, much less as a modern research facility. Plumbing and electricity were wholly inadequate, and not until I lost some samples that had taken weeks to accumulate because of an electrical failure was the electrical service in my lab brought to minimum standards. Particularly burdensome for my work was the lack of an ultracentrifuge that I needed for the subcellular fractionation essential to some of my projects. Penny had similar inadequacies at the Museum of Natural History. She did have a centrifuge but lacked a spectrophotometer, which I did have. So we traded the use of each other's equipment from time to time. More than once I walked the twenty blocks to the museum with tissue samples in an ice bucket for centrifugation, and she came down to use my spectrophotometer on several occasions. Soon after Bruce Gray, my first graduate student, had begun to work in my lab, he accidentally threw out some of her samples, wasting a week or more of her work. I don't think Bruce ever threw anything away in a lab again.

Those little disasters lay in the future that first Thanksgiving in New York. As I looked out on the marvelous view from my lab, then looked around the lab itself, I could already sense the tremendous risk I had taken in coming to New York for a job that I loved, but one that entailed such a time-consuming teaching load and subminimal conditions for any type of decent research. The struggle before me had begun to take form.

The period that Carol and I had together in New York between Thanksgiving and Christmas was all too brief. We flew back to Oklahoma and Texas to spend the holidays with our families, then flew to Chicago the day after Christmas to attend the annual meeting of the American Association for the Advancement of Science (AAAS). Ungar was there to present more exciting findings on the transfer of dark-avoidance behavior by a naturally occurring peptide from rat brain. The transfer paradigm was approaching its high-water mark.

18

More on Peptides

By the late 1960s, free amino acids and their derivatives were known to be present in the nervous system and functioning as neurotransmitters. The more obvious of these was glutamate, the most widespread excitatory neurotransmitter in the brain. Others include the tryptophan derivative serotonin and the tyrosine derivatives dopamine, epinephrine, and norepinephrine, which function both as neurohormones and neurotransmitters. Only three polypeptides, however, were known to be produced by nerve cells and play a role in neural function. They were oxytocin, vasopressin, and substance P.

Oxytocin is secreted from the posterior pituitary gland but is synthesized in the hypothalamus to which the gland is attached. It was originally discovered for its function in stimulating contraction of uterine muscle and the secretion of milk in mammals, though it has many different functions throughout the animal kingdom. It consists of nine amino acids, sequenced in 1953 by Vincent DuVigneaud, making it the first polypeptide for which the precise structure became known.[1] Soon thereafter, DuVigneaud sequenced a second polypeptide, vasopressin, also made in the hypothalamus and released from the posterior pituitary.[2] Substance P was not sequenced until 1971.

Discovery of the Pituitary Hormone-Releasing Factors

By 1970, it had become clear that several substances besides oxytocin and vasopressin with the characteristics of a peptide were being produced by the hypothalamus and transported through short blood vessels to the anterior pituitary gland where they caused the secretion of a number of hormones that regulate metabolism, growth, and reproduction. A frenetic race ensued among several laboratories to be the first to sequence what would become the third peptide known to be produced in the brain. Two laboratories—one headed by Roger Guillemin at the Baylor College of Medicine in Houston; the other by Andrew Schally at the Tulane School of Medicine in New Orleans— focused on the one releasing factor that appeared to be the smallest and therefore easiest to decipher. This was thyrotropin-releasing factor (later designated a hormone, or TRH). The two investigators, using somewhat different techniques, arrived at the same conclusion almost simultaneously that TRH is a tri-peptide (three amino acids) in late 1969 and early 1970.[3]

Roger Guillemin's lab was on a lower floor beneath Ungar's at the Baylor College of Medicine. The two men, with a shared heritage of classical physiological training and origins in France, knew and respected each other. Ungar claimed he was the one who first suggested to Guillemin that the hypothalamic releasing factors might be peptides. The key to Guilleman's success in deciphering the tripeptide structure of TRH was the new technology of mass spectrometry—a highly specialized technique that a twenty-eight-year-old scientist had been recruited, against his better judgment, to apply to the problem. The scientist was Dominic Desiderio, a brilliant young investigator who had taught himself the basics of chemistry from encyclopedias in the eighth grade and been the recipient of a Westinghouse Science Talent Scholarship to the University of Pittsburgh in his hometown. The reason Desiderio had been reluctant to take up the problem of the structure of TRH is that peptides are notoriously fragile and therefore difficult to analyze. But Guillemin had persuaded him to assume the challenge, and through persistent hard work, Guillemin's team came up with the answer. Thus, TRH became the third neuroactive peptide (after oxytocin and vasopressin) of known structure.

Dead on Arrival

At the same time that Guillemin was processing thousands of porcine hypothalami to extract enough TRH for analysis, Ungar was extracting peptides from hundreds of rat brains to obtain enough material for analysis of the factor he thought responsible for transferring dark-avoidance in rodents. One day, Desiderio attended a seminar by Ungar and learned of another peptide in need of structural analysis. Notwithstanding the tiny amounts of material that Ungar had obtained, Desiderio was fresh with enough confidence from his recent triumph with TRH to give Ungar's peptide a try. While this one turned out to be an even greater challenge than TRH had been, once again with persistence and hard work, Desiderio was able, by combining his mass spectrometric data with Ungar's conventional biochemical evidence, to come up with the proposed structure for a fifteen-amino acid structure (later reduced to thirteen amino acids) that Ungar decided to call scotophobin (from the Greek *scotos* for "dark" and *phobin* for "fear"). These were the results that Ungar had reported with gusto at the first ASN meeting in Albuquerque.

To get a peer-reviewed record of their achievement into the scientific literature, Ungar made the fateful decision to submit the paper to *Nature*; then *Nature* in turn made the fateful decision to solicit a review from a young expert on mass spectrometry at NIH named Walter Stewart. There ensued over a period lasting a year a highly unusual variation on the normal mechanism of peer review in which Stewart identified himself to Ungar (peer reviewers usually remain anonymous) and insisted on getting more details about both the behavioral and analytical chemical methodology. Ungar tried to respond, but never did so adequately in Stewart's view. When it became obvious that Ungar and Stewart were at an impasse, the editor of *Nature* took an even more radical step by publishing Ungar's four-page original paper[4] with an eight-page critique by Stewart,[5] back to back, followed by a brief rebuttal from Ungar and his coauthors.

Though the critique by Stewart was contrived, argumentative, and dismissive of fair context, it was technically logical and seemingly coherent. To a scientific public accustomed to skepticism over the behavioral transfer approach in general, it gave sufficient reason to dismiss the claim of a new, neuroactive peptide as unsubstantiated. To Ungar, the critique was beside the point. He and his colleagues had reported a peptide extracted from the brain with a specific sequence of amino acids. It was now up to the scientific community to test the precise function of what should have been recognized as the fourth neuroactive peptide of known chemical structure.[6] Instead, it was ignored and forgotten by the next and subsequent generations of neuroscientists, even as discovery of a multitude of neuroactive peptides was just around the corner.

The Bittersweet Taste of Hershey

For a time though, Ungar was a celebrity in the popular press. The second annual meeting of the ASN was held in Hershey, Pennsylvania, in March of 1971. Having submitted the paper to *Nature* just a month earlier, he was in a buoyant mood as chairman of a panel discussing the biochemical basis of memory. He had invited me to be on the panel, and I relished the opportunity to defend the transfer of learning approach, even as I had personally moved beyond the belief in any one-to-one correspondence between specific molecules and memory. I still believed that brain extracts were affecting behavior in recipients and felt strongly that further experimentation would eventually unravel the nature of the effect and thereby inform us of the precise nature of neuroactivity of endogenous peptides in the brain.

Embedded in these concepts was the question of how heterogeneous the brain really is at the chemical level. If different regions of the brain, and maybe even different cells have a unique chemical identity, then perhaps they have different antigenic properties. I, along with John

Hollis and Vincent St. Omer at Parsons, had decided on a simpleminded test of that possibility by screening for antigenic variations over different brain regions. One of my immunologically knowledgeable graduate student colleagues in Samson's lab, Stan Twomey, had endorsed this approach, and we had carried out some preliminary experiments that were inconclusive but I thought worthy of presentation at the Hershey meeting as an interesting approach to one of the fundamental questions in brain science. It was clearly a bridge too far for the immunologists in attendance, however, and they were critical of both my theory and methodology. My presentation was met with skepticism bordering on hostility. This, plus the tepid reaction to the position that Ungar and I had set forth in his panel discussion, didn't auger well for the rest of the meeting. My mood wasn't helped by another truly hostile reception to research presented by Samuel Bogoch, an eccentric scientist who was one of the few advocating, like Barondes and me, that glycoproteins and glycolipids might play some informational role in neural development or function. But Bogoch advocated this position in nonrefereed publications, and his techniques were highly questionable, including by me. Unfortunately, he would later invoke my name in support of his cause, and given his reputation, this did me no favors.

All in all, the Hershey meeting gave me a good bit to be upset about. But I also have fond memories of that meeting because of a number of new friends that I made there. I met Richard Quarles and Cara-Lynn Schendgrund, both interested in glycoproteins and glycolipids, for the first time. Arnold Golub, who had conducted some of the most ingenious of the transfer experiments, was there, and he *had* understood what Ungar and I were getting at. Golub by then had moved to the Eunice Kennedy Shriver Center for Mental Retardation in Waltham, Massachusetts; and the head of the biochemistry division there, Robert McCluer, was also in Hershey. I needed no introduction to McCluer—as one of the few biochemists proclaiming the importance of gangliosides for some time, he had done much of the important early research on their chemistry. The meeting lasted all week, and there wasn't a lot of diversion in Hershey; so McCluer, Golub, and I spent a number of hours over drinks, talking about transfer experiments, gangliosides, immunological approaches, and the scientific study of learning and memory. Bob McCluer, who would become a fast friend and an important figure in my later career, displayed from the beginning of our relationship a quality that I greatly admired: a broad tolerance for novel approaches and unconventional ideas, despite the straightforward, conventional approaches of his own research.

The Hershey meeting had a couple of other redeeming features. Fred and I had a good visit, and Dianna and I got to catch each other up on gossip again for the first time since my leaving Kansas. I also enjoyed my role as the one colleague that Georges Ungar could count on to be receptive and respectful, if not always agreeable. But overall, I was more than happy to return to New York, where Carol had finally arrived from Kansas. She still had a dissertation to write but really *had* managed to finally finish that "one last experiment."

[1]   Du Vigneaud, V., C. Ressler, and S. Trippett. 1953. The sequence of amino acids in oxytocin, with a proposal for the structure of oxytocin. *J Biol Chem* 205: 949–57.

[2]   Du Vigneaud received the Nobel Prize in 1955 for sequencing these two compounds.

[3]   Burgus, R, T. F. Dunn, D. Desiderio, and R. Guillemin. 1969. (Molecular structure of the hypothalamic hypophysiotropic TRF factor of ovine origin: mass spectrometry demonstration of the PCA-His-Pro-NH2 sequence). *C R Acad Sci Hebd Seances Acad Sci D* 269: 1870–3; Nair, R. M., J. F. Barrett, C. Y. Bowers, and A. V. Schally. 1970. Structure of porcine thyrotropin releasing hormone. *Biochemistry* 9: 1103–6. Guillemin and Schally shared the Nobel Prize in 1977 for their discovery.

[4]   Ungar, G., D. M. Desiderio, and W. Parr. 1972. Isolation, identification and synthesis of a specific-behavior-inducing brain peptide. *Nature* 238: 198–202.

[5]   Stewart, W. W. 1972. Comments on the chemistry of scotophobin. *Nature* 238: 202–10.

[6]   Irwin, L. N. 2007. *Scotophobin: Darkness at the Dawn of the Search for Memory Molecules.* Lanham, MD: Hamilton Books. The discovery of the releasing hormones, Desiderio's role and background, and the bizarre nature of the publication of the structure of Scotophobin are covered on pages 129–132.

19

Sadness and Restoration

On June 1, 1971, my father died in San Antonio from the complications of congestive heart failure. He had been in failing health for some time and was in the hospital recovering from surgery. I knew he was very sick, but the suddenness of his death caught me by surprise and was a stunning blow. The previous autumn, he had come to New York for the first time, already weakened by heart failure and almost blind from a congenital hereditary condition; but he had managed to find his way around the city on his own, and that had been immensely satisfying to him. He took the greatest pleasure in proclaiming to waitress and customers alike at a neighborhood diner that he was up from Texas to visit his son and buy him a steak—both the proclamation and the steak being sort of a ritual for Texans (the only people in the world who brag about where they're from in the first thirty seconds of small talk).

The day he died was bittersweet in the extreme because only that morning, Carol had received confirmation that she was pregnant. I had told my mother and sister to tell my father before we got it confirmed, if he seemed to be taking a turn for the worse. My sister exercised her judgment to do so just shortly before he was struck by a pulmonary embolism that night. Hearing that Carol and I were expecting a child was apparently the last news he heard of his oldest son.

A Disappointing Site Visit

I returned to New York from the funeral deeply depressed but with no time to dwell on my feelings. Awaiting me was a letter from NIH, which I took at the time to be good news. In January, I had submitted a grant proposal for research on glycoconjugates (glycoproteins and gangliosides) in the developing retinotectal system of the chick embryo, and the study section to which it had been referred had found it of sufficient interest to want to make a site visit to my lab to obtain more information. *Past the first hurdle,* I thought. So Bruce Gray, my new graduate student, and I swung into action immediately to generate as much data as possible prior to the site visit, which was set for the end of July.

My proposal was aimed at determining whether a spatial gradient exists within the region of the optic tectum to which the nerve fibers from the retina are connected. Since it was known that a nerve cell at a particular point in the retina projects to a highly specific point in the tectum, it had long been assumed that a chemical code of some sort tags each point in the tectum with a molecular marker that enables the ingrowing fiber from the retinal cell to recognize its proper destination. I agreed with Barondes that the glycoconjugates, because of their location at the cell surface and their high information content, were likely repositories of such a chemical code if it existed. And if it did exist, the glycoconjugate profile of one point in the tectum should differ from that at another point, just as the house numbers at one end of a street differ from those at the other end. There was widespread agreement on this theoretical prediction. What I proposed to do that was new and different was to divide the tectum into a number of tiny different blocks of tissue and directly analyze the chemical composition of the glycoconjugates of each individual block. So Bruce Gray and I set about learning how to dissect the optic tectum—a piece of tissue at early stages of development about the size of a single grain of rice—into nine blocks. Using microsurgical techniques I had learned in graduate school, Bruce and I, with practice, got to where we could perform this delicate operation under a dissecting microscope with enough confidence to demonstrate our skill to the site reviewers. This seemed particularly important because the letter from NIH had indicated an interest by the site visitors in seeing our technique with the embryos. By contrast, the chemical procedures were inelegant and simple but still promised, in my view, to yield useful information. They certainly would yield information that no one had obtained before. So our days in June and July were long and hard as we prepared for the arrival of our judges.

My spirits were buoyed in early July when I received word of my first external grant (since the NSF graduate fellowship)—a five-thousand-dollar award from the Merck Company Foundation. Of the forty grants for faculty development made nationwide, mine was among only two or three made to a faculty member not at a medical school. My dean knew the people at Merck who made the awards, and this probably helped me; but I allowed myself the luxury of assuming that the award was based on merit. So far, so good.

As the date of the site visit drew nearer, Bruce and I committed ourselves to heavy-duty homework. I thought up 105 questions that I might be asked and rehearsed all the answers. We practiced the dissections time and again and made sure we would have fresh embryos available on the appointed day. Bruce scheduled an experiment (light work, not requiring much concentration) the day of the visit to show that the lab was busy at all times. The night before, I got a good night's sleep. We were as ready as we were ever going to be.

The site visitors were Marcus Singer and Irving Lieberman, both noted developmental biologists with strong biochemical orientations. They arrived a little early on July 27, so they made their way through the archaic building to my fourth floor lab without the benefit of an escort. *Not a good way to begin*, I thought. They then proceeded to grill me for the better part of four hours, asking mostly the questions I had anticipated. I gave my well-rehearsed answers with as much conviction as possible, but I had the feeling they were skeptical from the start. If I found a chemical gradient across subregions of the tectum, how would I know it was information that the developing system really uses? If I didn't find a gradient, how would I know that one didn't exist anyway among components I had failed to detect? What about all the cells in the vicinity of the retinotectal connections, wouldn't their chemistry swamp out the tiny minority of molecules coding for the locus-specific information? And how could the chemistry, as simplistic as I had proposed it, do justice to the true complexity of the molecules I was trying to study? In one way or another, they were telling me through their questions that it didn't matter what my findings revealed—the results wouldn't mean anything anyway.

The dissection demonstration turned out to be anti-climactic. They even offered to let us skip it (perhaps they had heard enough), but we insisted, and Bruce did a fine job. They were cordial and respectful throughout and were generous in thanking us for our time. After they left, we retired to the neighborhood Steak & Ale, where Carol joined us for a "celebration." I was glad the site visit was over, and I didn't think I had disgraced myself, but I didn't have a very good feeling about the eventual outcome.

When the letter of rejection finally came, it cited what I had feared: their failure to believe that any result could be interpreted in a useful way. In the hindsight of time, I now see that it was the significance of the research that was oversold. Like Ungar with scotophobin, I was implying that my results might have greater meaning than they possibly could. But like scotophobin, the research I had proposed to do was breaking new ground. No one had looked at any glycoconjugate chemistry in the retinotectal system with any anatomical resolution at all. There had to be merit in this initiative, however crude, just as there was surely merit in characterizing neuroactive peptides, whatever their exact role turned out to be, as subsequent history would show. It was just a matter of excessive expectations and inadequate technology.

A Brilliant Dissertation Defense

Immediately upon Carol's arrival in New York in February, she had begun writing her dissertation. All day every day she worked on it, interrupted only by the brief excursions to join me in Hershey, visit friends in Baltimore, and for the trip to Texas for my father's funeral.

She finished writing the week of my site visit at the end of July and flew to Kansas two weeks later for her oral defense. She called me from Lawrence, depressed after rehearsing her oral presentation with Dick Himes on Sunday night. It had not gone well, and she was not feeling confident. I assured her that the real thing would be great and told her to think positively. Tuesday night, August 17, she called in a mood of euphoria. Her presentation had indeed been great, her answers outstanding, and her overall defense of the dissertation brilliant. Dick Himes told her it was the best-written dissertation he had ever directed, and her committee voted to give her the rare grade of honors.

With my site visit behind me and Carol's dissertation out of the way, we decided that a trip to the wilderness of Maine would be in order. As we drove into the lovely woods of Baxter State Park in northern Maine, we finally began to unwind from the strain and sadness of the summer. With time to be introspective, I let the full force of depression from my father's death wash through me. Then, in the tranquility of that lovely place, I felt it ever so slowly begin to recede. I had taken data from some of the summer's research with me to work on, as calculating data is an aspect of research that I always found enjoyable and relaxing, even on vacations. Carol didn't share my enthusiasm for working in such a setting, so she was content to curl up with a mystery novel while moose browsed in the meadow and loons took off and landed in the stream beyond our campsite. On a couple of occasions, a bear cub poked its head around our lean-to and one night made off with my leftover campfire stew. By this time five-months pregnant, Carol was not in the mood for surprises, so the bear made for more wilderness than Carol really wanted, but we stayed our allotted three days without incident. I climbed the second highest mountain in the park and realized I was the furthest north I had ever been in my life. The cold rainy wind that whipped my face as I reached the summit made me think of the other ways in which all the new experiences in my life since leaving Kansas had been less than a piece of cake. I was glad to be where I was—the scenery was worth it. But there was no one else up there sharing the view.

20

More Meetings, a Birth, and a Spot on the Nightly News

A new batch of freshmen arrived in September at the College of Pharmaceutical Sciences, Columbia University, even as the institution was beginning to face up to a grave fiscal predicament. The second oldest pharmacy school in the country, it had a proud and accomplished history; but it no longer could compete with the public university system in New York, which charged a lot lower tuition for an education of at least comparable quality. The affiliation with Columbia, it turned out, was hardly more than in name only, and Columbia University was not about to absorb the albatross of the pharmacy school. Faculty salaries had been frozen the previous spring in the face of a half million dollar deficit. The students and their families, of course, had not been informed of the fragile condition of the college. They arrived as fresh and eager as the eighteen-year-olds that had arrived every September before them—but perhaps with even more seriousness of purpose, as the grinding war in Vietnam hung over them and, for many of them, drove the wedge deeper between them and their parents.

Getting to Class on a Crippled Bus

Life seemed more somber to me too, as it must to anyone after losing a parent. But one of the great perks of the academic life is that every September offers new hope and a fresh start. Freshmen arrive with enthusiasm, and they never grow older, year after year, giving their teachers a sense of renewal. Infused with these thoughts, I threw myself into the task of impressing the breadth and beauty of modem biology on a fresh batch of future pharmacists.

Bruce and I plunged forward with our work, hopeful that our biochemical studies of the retinotectal system would turn up information useful to understanding the formation of specific nerve connections, notwithstanding the doubts of our critics. And three undergraduates in my lab—Judy Mancini, Donna Hills, and Bram Trauner—were turning out some pretty good work on the properties of rat brain sialidase, an enzyme that resculptures gangliosides and certain glycoprotein molecules. However, working with inadequate equipment in such poor facilities was a constant struggle. And as the dire fiscal state of the institution became increasingly clear, any hope for renovations or capital expenditures to improve my research conditions faded. Nonetheless, we persisted, and it looked like we would have some hard data worth publishing before long.

In addition to a reprise of my general biology course, I offered a graduate seminar in neurobiology for the first time. Since several of the graduate students were clustered near Bruce and and his wife Sally's upper West Side apartment on Riverside Drive, we decided to hold our sessions there. Once a week after dinner, I would catch a bus for the twenty-minute ride up Riverside Drive. One evening, the engine at the back of the bus seemed about to catch on fire. The driver pulled over, checked the damage, and told us that the engine was smoking badly but that he thought we could keep going if we wanted to. He asked for a show of hands from those who wanted to evacuate the bus. By an overwhelming majority, we voted to keep going. I got off at the appropriate point, and the bus drove on, trailing an ever-denser cloud of smoke.

Another Inaugural Meeting

From many different directions, the neurosciences were converging into a coherent and distinctive discipline, as formation of the Society for Neuroscience about two years earlier had demonstrated. So clearly it was time for that society to start meeting annually, as the ASN and ISN had already begun to do. Washington DC was declared the site for the first meeting; so Bruce, Sally, and I with Carol piled into his car with enough luggage for a month's safari for the drive to the week-long gathering. Like the first meeting of the ASN, this one was occasioned by a lively spirit and the great anticipation that comes with being at the cutting edge of a new and worthwhile venture. The founders had made the crucial decision to make the society reasonably inclusive and non-elitist, hence easy to join, and apparently had hoped that it would be a forum for testing new ideas, including new ways of holding scientific meetings. There were 1,396 attendees at that first meeting, so it looked in the beginning like such objectives could in fact be achieved. As membership ballooned by six fold within the decade, the aim of inclusiveness was realized, but the hope that this society's meetings would be much different from those of any other society were dead by the end of the following year, as we shall see.

From the stimulation of the Washington meeting, I returned to the grim reality of the situation at the College of Pharmaceutical Sciences. Factions began to develop within the faculty, as some members with long-standing grudges against the dean tried to use the situation to oust him. Morale plummeted and students were drawn into the vortex of anxiety about the college's future. Bruce became a leader of the graduate students and gave a couple of impassioned speeches, though to what end, I don't remember. The fact remained that there really wasn't much anyone could do.

Philip Merker, the chairman of my department, finally spelled out for me in unambiguous terms what I should have figured out sooner myself. My job was in severe jeopardy, and I should start looking for a new position as soon as possible. I had put off facing this reality because I had enjoyed living in New York so much and, despite the heavy teaching load, was living my dream of college teaching with a great deal of satisfaction. On the other hand, getting research done had been a real struggle; I had felt isolated from others working in my field, and the academic and collegial atmosphere had deteriorated to an abysmal level. My search for a new job, reluctantly, began in earnest.

The Joy of a Life-Changing Event

The full impact of these problems was just coming into focus as Carol was eight months pregnant. During her first trimester, she had been busy writing her dissertation. During her second trimester, she defended her dissertation, took the trip to Maine, and started a postdoctoral appointment at the Albert Einstein College of Medicine in the Bronx. Not until her third trimester did both of us begin to think a lot about the radical change that was about to come over our lives. We worked quickly to make up for lost ground, reading books on pregnancy (Guttmacher) and childcare (Spock), like the other members of our generation who didn't have the benefit of an extended family nearby to pass along knowledge in the customary oral tradition. We took a few childbirth classes, though not enough to learn to do it right. By the first week in December, Carol was persistently uncomfortable; and with the ill fate of the college as well as the impending change in the status of my family much on my mind, I was out of sorts much of the time and hardly in a holiday spirit.

On the twenty-first of December, Carol went with me to my nearby office to write part of the paper she was preparing for publication from her dissertation and, by the end of the day, was beginning to have mild contractions. I fixed dinner for myself, but she had only a Jell-O salad, in case she really were going into labor. We watched television specials on a day in the life of the presidency (Nixon) and on the origins of the Vietnam War, neither topic being particularly pleasant but both serving as a diversion as her contractions increased in frequency. By midnight, we were ordered to the hospital. It took us ten minutes on the corner of Broadway and West Seventieth Street to catch a taxi, as somebody jumped into one ahead of us when I—with a suitcase, load of pillows, and obviously pregnant wife—didn't reach for the door handle fast enough. Nonetheless, we checked in by twelve thirty and were into the labor routine by 2:00 a.m. The doctor had predicted delivery by 7:00 a.m., but problems developed, including an infusion line that was botched by the nurses. With all the commotion of nurses and doctors in and out of the room, the teamwork between Carol and me broke down, and feeling pretty useless, I finally left about 8:00 a.m. It turned out that the baby was facing the wrong way, and that took some manipulation; but our son, Anthony, was finally born without incident at 9:38.

Though I felt I had acquitted myself less than admirably during the birth process, I was much relieved to see Carol on the road to recovery a few hours later and apparently bearing no grudges over the ordeal through which she had just passed. I got back to our apartment and called various members of the family with the news. Then, suddenly and inexplicably, I got the Christmas spirit. I tipped the desk attendant and doorman of our building more than I had ever thought about and, for a time, walked the streets of New York smiling at strangers as though I were back in Houston. It was an unnerving feeling, and it passed, but it helped get me through Christmas with Carol and Anthony still in the hospital. On New Year's Eve, I took the subway to Times Square shortly before midnight to be a part of the world's largest annual street brawl. As January 1, 1972, arrived in the flush of personal excitement, I managed to temporarily ignore the oppressive clouds on my professional horizon.

Sticking to the Program

Symptomatic of the way the year was to go, I came down with a bad case of the flu about ten days into 1972 and had to cancel a lecture for the first time in my academic career. Soon I was back in the lab, however, working hard on the research on sialidase enzyme activity with Judy and Donna and on glycoprotein and ganglioside changes in the retinotectal system with Bruce. As the financial situation of the college had deteriorated, and as faculty and student morale had plunged with it, sustaining a normal academic and scholarly atmosphere became very difficult. I insisted, however, that my lab would stick to the appointed schedule of research, even as the physical and social infrastructure crumbled around us. By contrast, most of my faculty colleagues were intensely engaged in activities designed to save the college, oust the dean, or both—neither of which, I had come to believe, was likely to succeed. One of their main contentions was that the demise of the college would mean the loss of a valued research and teaching institution, though most of them were doing little teaching and no research by then. To publicize the plight of the college, they managed to get a local television crew on the premises one day. This occasioned great excitement, though characteristically, my students and I stayed clear of the hubbub, working in my lab. The crew asked one of the professors to pose in a lab as though he were doing research. Then, at the height of the effort to contrive such a scene, someone mentioned that there were people doing real research in another lab upstairs, whereupon the reporters and cameramen abandoned the contrived scene and found their way to my lab instead. So it was my students that ended up on the evening news, and deservedly so.

Alone and Neglected in Purchase, New York

Phil Merker, the departmental chair who had hired me, felt badly about the fact that I was clearly going to be the first casualty of the institution's demise. On my behalf, therefore, he tapped what contacts he could. One of them was John Hildebrand, a geneticist from Columbia University who had recently moved to the new State University of New York (SUNY) campus at Purchase, just north of New York City. He offered me an interview, so I took the commuter train up to Purchase on the appointed day, then waited at the train station where I had been told I would be picked up. When no one came after forty-five minutes, I called Hildebrand's office to make sure I had come on the right day. Unfortunately, I had, which meant I had been forgotten. I didn't go into that interview with high hopes that a job would materialize, and on my way home, I realized that there probably had never been an opening to begin with.

On July 28, *Nature* published the paper by Ungar, Desiderio, and Parr on scotophobin, along with Walter Stewart's rebuttal. Involved as I was at the time in adjusting to parenthood and seeking a new job, I was unaware of the paper, though I knew of the work. Even less was I aware of the real news of 1972 in neuroscience—the train of events then underway that succeeded where scotophobin failed in convincing the scientific community of the pervasive importance of brain peptides.

21

End of an Era

As 1972 was coming to an end, my professional life was going from bad to worse, and the body politic of the nation seemed to be doing the same. The College of Pharmaceutical Sciences had become a totally demoralized institution with no hope of salvation. I had accepted the fact that my position there was going to end, probably sooner than later, just as I got back two manuscripts that I had written in the summer— one with Bruce Gray on sialic acid changes in the developing chick embryo, the other with Judy Mancini and Donna Hills on sialidase activity in the rat—both of them rejected for publication.

The presidential campaign lurched toward its inevitable dismal end, with Secretary of the Treasury John Connally making television commercials on behalf of Democrats for Nixon, and Secretary of State Henry Kissinger hinting that peace was at hand in Vietnam despite all evidence to the contrary. Distressed by Nixon's impending reelection and depressed by the two rejected papers, I at least had the second annual meeting of the Society for Neuroscience in Houston to look forward to. Going back to Houston for any reason should have lifted my spirits, but even that turned out to have its share of sour notes.

Outraged in Houston

The Houston meeting was a disappointment for a couple of reasons. Georges Ungar, in offering Houston as the host city the previous year, had envisioned it as a showcase for his discovery of scotophobin; but by the fall of 1972, it was obvious that the neuroscientific community was reacting with great skepticism toward his claims. For the society itself, that second annual meeting was pivotal in determining whether the society would develop a creative, new approach to interactive scientific communication at a meeting or would follow the conventional pattern of isolated paper presentations characteristic of all scientific societies once they exceed a few hundred members. At that point, the Society for Neuroscience was still small enough to experiment with innovative approaches and seemed intent on doing so.

I had submitted an abstract on conceptual and methodological impediments to the study of memory mechanisms. With surprise and pleasure, I received a letter from Arthur Ward, a well-known neurophysiologist at the University of Washington, informing me that my paper had been selected as one of a small number to be presented in a special session where an innovative format other than the conventional ten-minute oral presentation would be tried. Perhaps a brief introduction of basic concepts, followed by a round-robin discussion or debate or some other novel format to be decided by the participants, would be appropriate. He was thus soliciting my ideas about the form the session should take. I welcomed this overture since (1) the paper I intended to present did indeed lend itself well to a discussion/debate format, and (2) I thoroughly agreed that the society should be innovative in exploring new modes of scientific communication while it was still young and visionary enough to do so.

I spent the better part of a weekend drafting a long letter to Ward, outlining a number of ideas about the format that we might adopt and soliciting his reaction to those ideas. I waited in vain for feedback. As the meeting date approached, it became clear that the best we could hope for was a final decision made presumably by him as to what the format would be, but by the time I had to leave for Houston, not even that had arrived. Thus, I showed up at the session (discovering that the other participants did also) with no notion whatsoever of how the session would be conducted. Since we had clearly been led to believe that conventional presentations would *not* be in order, that was the one format for which I had not prepared. I met Ward for the first time just minutes before the start of the session. He didn't know me, of course, nor did he appear to care; and he offered no apology for his failure to respond to the ideas that I, and apparently others, had offered at his solicitation. Instead, he proceeded to open the session with the announcement that, in his judgment, the best way to proceed would be with each speaker making a standard ten-minute presentation to be followed by discussion if there were to be any time remaining.

To say that I was furious is an understatement. Unprepared for an extemporaneous ten-minute speech and seething at the thought of all the wasted time I had spent agonizing over creative new approaches to scientific communication, I somehow managed to get up and talk for several minutes with nothing worse than a brief but biting apology for my misunderstanding of what the format was to be. Fortunately, I had prepared a novel slide—a pictorial representation of a dozen different proposed functions for glycoconjugates (glycoproteins and gangliosides)—to illustrate the need for us to be more creative and expansive in our view of the cellular mechanisms that might be involved in the storage of neural information. It was brightly color-coded and magnificently gaudy, designed to grab attention and make a point. Whether it succeeded or not, I don't know, as predictably, we didn't end up with much time for interactive discussion about one another's presentations. But it did give me a visual focus for a quickly contrived extemporaneous presentation.

There had, of course, been no misunderstanding about the format— only deception from the chairman about what it would be. In retrospect I have been unable to bring myself to believe that Arthur Ward, a respected and undoubtedly sincere scientist, deliberately intended to make a mess of the program. It is also clear that my hostile feelings about the episode outweigh the dimensions of the crime. At the time, though, it seemed so symptomatic of some of the characteristics of institutional science that I had come to resent—from the reliance on established authorities and conventional ways of thinking to the exclusion of more creative ideas propounded by newer, as yet unknown voices. I'm sure that what happened was that the society's founders genuinely wanted to try something different—this would have been characteristic of Ralph Gerard—but as usual, they turned to an established figure to carry out the experiment, and he was simply too busy or too set in his ways to make anything of the opportunity.

After the session, Fred Samson and I retired to a local bar for cocktails, as I poured out all the venom that had built up in my mind over this; and he listened with the same tolerant sympathy for the wayward wild-eyed views of his former student that I had, by then, come to expect and appreciate.

I returned to New York from Houston—upset by the seminar fiasco and depressed by the obvious failure of Ungar's work to have the impact that I thought it deserved—back to the morass of a dying institution and a litany of professional setbacks. It was in this mood that I wrote a plaintive letter to Jerry Mitchell in Detroit. In September, he had accepted an appointment as assistant professor of anatomy at the Wayne State University School of Medicine. I outlined my woes and appealed to him for any help or suggestions that could lead me out of the darkness into which I had fallen.

Then I retreated to the sanctuary of my office, determined that my two rejected papers would someday see the light of day. Both were rewritten and resubmitted. The sialidase enzyme paper was frankly of marginal importance, but Judy especially had worked hard on it; I wanted her and Donna to get a publication out of their efforts, and I needed the publications to remain viable. The developmental paper with Bruce Gray was more important, if less substantive. Speculations were becoming more common in the literature about the possible role of glycoproteins and glycolipids in the development of the nervous system, without the benefit of any data to support the speculations. Our skimpy paper would at least bring some facts to bear on the problem.

While working on the reclamation of those two papers, Jerry called me from Detroit and told me to send him my résumé, which he would circulate through several departments at Wayne State. He knew of no openings there, and I wasn't particularly hopeful, but his response to my signal of distress was a kind gesture that lifted my spirits. An updated résumé was dispatched to Detroit as we made plans to go to Texas for the Christmas holidays.

Grinding and Binding Gets Going

The real news of 1972, which made my personal melodrama pale into insignificance and eventually even banish scotophobin from the front pages of neuroscience, was another frenetic race like the one to discover the first hypothalamic releasing factor—this one in pursuit of the opiate receptor, which ironically circled back to Ungar's original search for the mechanism of morphine tolerance. By the early 1970s, it was obvious that receptors must exist in the brain that, when attached by narcotic opiates like morphine, would reduce pain. However, animals quickly develop tolerance to opiates like morphine and its synthetic mimic, heroin. Thereupon, withdrawing the drug generates debilitating physiological and psychological consequences that spur the desire to take in higher doses of the drug—the essence of addiction. It was the alarming rate of heroin addiction—in some estimates as high as 30 percent—among soldiers returning from Vietnam that spurred Nixon to declare a war on heroin as part of his "law and order" agenda.[1]

One of the first recipients of "war on heroin" money was a leading pharmacologist at Stanford, Avram Goldstein. Author of a definitive textbook on *Principles of Drug Action*, Goldstein was anxious to return to research after years of administration. With a $400,000 grant, he set up an addiction research lab and proceeded to attack the problem of the receptor in the brain for opiates like heroin. In 1971, he had described a technique that in principle could detect the presence of receptors to any chemical substance. Homogenates of tissue suspected of having receptors to a particular chemical compound could be ground up in a homogenizer, then incubated with a radioactively labelled ligand (a specific compound) suspected of binding to the presumed receptor.

When unbound ligands were rinsed out of the homogenate at the end of the incubation period, the label that remained stuck to the particles of tissue would indicate the presence of receptors for the ligand bearing the label. Using this technique, he had found that levorphanol, a chemical variant of morphine, would bind to homogenized brains from mice, but only at a disappointingly low level about 2 percent above background.[2]

Lars Terenius, a young assistant professor at the University of Uppsala in Sweden, was well familiar with the receptor-binding technique, having used it to study intracellular receptors to the female hormone estrogen. He had improved upon Goldstein's method by using ligands with much higher specific activity of the label, meaning that a higher proportion of the ligand molecules actually carried the radioactive tag. This significantly increased the sensitivity of the assay. With this improvement, Terenius had found specific binding of opiate drugs and submitted a paper on his results in early November 1972 to a Scandinavian journal,[3] where it languished *in press* while two other labs in the United States closed in on the problem.

Solomon Snyder at the Johns Hopkins School of Medicine was a well-established neuropharmacologist with an interest in the chemical basis of mental illness. He heard Goldstein describe his receptor-binding technique at a talk in the summer of 1971. He didn't have a very high opinion of opiate research, but the availability of "war on heroin" funding made the problem attractive enough for him to assign one of his new graduate students the task of detecting a higher level of binding than Goldstein had achieved. Candace Pert, an eager, vivacious, and imaginative graduate of Bryn Mawr was the student recruited for the task. By good fortune, Snyder's lab was next to that of Pedro Cuatrecasas, famous for his detection of insulin receptors. He had invented a multiple manifold machine consisting of a large number of wells for holding the homogenates as they incubated on a glass fiber filter with the ligand. By vacuum filtration, each sample could then be quickly rinsed, enabling a large number of samples to be processed simply and in parallel.

Pert threw herself into the project with characteristic vigor, but labored throughout the summer of 1972 without success. Finally, in mid-September, she tried labeled naloxone, a powerful opiate antagonist, and that gave spectacular results. She then studied a number of other analgesic drugs and found that their binding affinity correlated nicely with their pharmacological potency. Pert and Snyder announced their findings with great fanfare at a news conference four days before the experiments were published in *Science* on March 9, 1973.[4]

Eric Simon, a professor at the New York University School of Medicine, had conceived the idea of testing for opiate binding using radiolabeled nalorphine as far back as 1965 but had abandoned the effort when it didn't work. Goldstein's paper in 1971 had rekindled his interest, however, and he returned to the problem, this time using etorphine, a narcotic thousands of times more potent than morphine. Just as he was getting strong evidence for etorphine binding in brain homogenates, the paper by Pert and Snyder came out in March. Dismayed by the fact that he hadn't gotten into print with his evidence first, he was nonetheless scheduled to present his paper orally at a meeting the following month. The printed version came out in July.[5] Overall, it's fair to say that strong evidence for the presence of an opiate receptor in the brain was acquired essentially simultaneously by the labs of Terenius, Snyder, and Simon, prefaced by earlier and much weaker evidence from Goldstein.

In Aberdeen, Scotland, Hans Kosterlitz and his postdoc, John Hughes, were already working on the next logical step: discovery of the endogenous substance for which the receptor must exist. Kosterlitz, like Ungar, had been trained in the classical European tradition of physiology and medicine, where bioassays were commonly used to probe for physiologically active natural substances. Kosterlitz was in fact using contraction of the guinea pig ileum—the same preparation Ungar had used in his early studies of antihistamines—to study the narcotic effect of candidate opioids. Since the chemical structure of morphine doesn't resemble a peptide at all, they had discounted the possibility that the endogenous ligand for the receptor could be a peptide. But Terenius, who had been invited by Kosterlitz to Aberdeen for a brief period, believed on the basis of his own search for an endogenous opiate that one existed and had the properties of a peptide. Terenius persuaded Kosterlitz to try incubating his preparations with peptidase, an enzyme that breaks apart polypeptides, and the resulting preparation showed total inactivity as a narcotic in the guinea pig ileum assay. From that day on, they decided they were in a hunt for a peptide. But it would take another year and a half before they found it.

"You'll Need a Property Voucher and an Ambulance to Bellevue"

Just at the moment that our futures appeared most bleak back in New York, Carol and I received a sign that a better day was coming. To distract us from our troubles, we decided to go to Lincoln Center one evening for a performance of *La Bohème*. This was hardly an inspired choice if the point was to lift our spirits, but the opera did remind us that the human spirit can suffer greater privations than a lousy meeting and two papers rejected for publication. So we stopped at the Monks Inn, our favorite restaurant near Lincoln Center, to enjoy a late supper with wine and to count what blessings we could. I don't remember how many blessings we tallied, but I do remember that we had a lot of wine because I was very lightheaded as we started our walk home. Just as we passed the front door of the Ginger Man, a touristy West Side restaurant known for its colorful and often affluent clientele, I spied what appeared to be a roll of money lying on the sidewalk beside a limousine a half a block long. I picked it up and showed it to Carol, wondering, *Why would anybody be carrying around a wad of play money like this?*

For some reason, it struck me as very funny (things will, in that state of mind), but as I started to laugh about it, Carol turned a more sober eye to the wad of bills she had taken from my hand, and said, "Louis, this isn't play money; these are real bills!" And so they seemed to be. I began to sober up quickly, but not fast enough to prevent me from turning impulsively to the chauffeur standing by the door of the limousine and asking him, "Is this yours?"

For a very long moment, he stared at the bills without comment, then mumbled something to the effect that he didn't know anything about them, but if he were us, he would be walking on down the street right away. Carol thanked him, grabbed my arm, and set the pace for a

brisk walk back to our apartment. Once safely inside, we examined the
roll in detail and concluded that, indeed, the bills were either genuine or
very good counterfeit specimens, adding to a total of seventeen hundred
and sixty dollars.

We contemplated our course of action as I continued to sober at an
impressive rate. Should we turn in the money, fearing that it might be
stolen or counterfeit? Should we chalk it up to good fortune, spending
the fifty- and hundred-dollar bills here and there, now and then, as
inconspicuously as possible? We decided to sleep on it, literally. Carol
placed the roll of bills under her pillow, and we went to sleep thinking
not at all about our tribulations for the first night in a long time.

In the fully sober light of the next morning, we decided to turn the
money in—our sense of honor asserting itself, not to mention the fear
of getting caught passing stolen bills. So Carol buckled Anthony into
his stroller and took herself, her infant son, and her wad of $1,760 to
the nearest police precinct a couple of blocks away. Within minutes of
her arrival, it became clear that a woman with a baby turning in a large
amount of cash was not an everyday occurrence in Manhattan. The desk
sergeant looked at her uncomprehendingly when she explained that she
had found $1,760 lying on the sidewalk and wanted to know what to
do with it. Finally, he pulled himself together enough to ask her, "Well,
did you report it?" Carol, under the impression that this was what she
was trying to do, answered, "Not yet." Confusion reigned supreme. The
receptionist sent for his commanding officer.

The lieutenant, who was a bit quicker mentally than his underling,
proceeded to examine Carol, then the money, then Carol again, with
bemused wonder. Word spread through the station quickly: a woman
had found a couple of thousand dollars in cash on the sidewalk and
had come to turn it in. Offices emptied, desks were abandoned, and
paperwork languished while New York's finest gathered to witness
the strange event unfold. Such was the rarity of someone turning in,
as opposed to reporting the loss of, property that the lieutenant didn't
know a form for such an eventuality existed.

"Hey, Joe," he yelled back to his captain. "This woman wants to turn in seventeen hundred and sixty dollars in cash that she found on the sidewalk. What do I do?"

There was a momentary pause, then a droll voice from a rear office intoned. "Well, you'll need to fill out a property voucher . . . then call an ambulance to take that lady to Bellevue!"

A carnival atmosphere ensued. Officers milled around, quizzing Carol for details, talking baby talk to Anthony, peering at the money, wondering to themselves whether they were witnesses to high drama or an ingenious hoax. Undoubtedly, an armed robbery across the street would have gone unnoticed and certainly would have elicited less surprise than Carol's appearance with the money that morning. At one point, a sympathetic patrolman offered her a cigarette, which she politely declined.

"No, thank you. I don't smoke."

"Jeez, lady," retorted the officer, "don't you do anything wrong?"

It turned out that there was a law on the books for such a situation. Found money had to be impounded for a period of time—in the case of $1,000 to $2,000, for twelve months. If no one claimed the money by the end of the specified waiting period, the finders got to keep it. No one did claim the money, and a year later, we got a check for $1,760 from the New York City Police Department.

Our years in New York City were an incredible mixture of good fortune and hard times. In November of 1972, we were reaching the nadir of our experience there. But the night we found a roll of money on the sidewalk in front of the Ginger Man marked a turn for the better in our mental as well as our material fortune. Within a month, the war in Vietnam would take a dreadful turn, but on a personal level, our lives had rounded the corner for the better.

Death of a Giant

The spring offensive launched by North Vietnam in 1972 had been broad, brutal, and very costly on both sides. It served to secure more territory for the communists but elicited the mining of Haiphong Harbor and renewed bombing of North Vietnam. Secret negotiations in Paris between Henry Kissinger and Le Duc Tho began to make real progress in September, and by early October, the outlines of a peace accord had been agreed to.

Kissinger announced at a White House press conference on October 26 that "peace is at hand," even though South Vietnam's Thieu regime was in fact still blocking a final agreement. But coming just twelve days before the election, Kissinger's words had the intended effect of leading the public to believe that an end to the war was imminent, thereby removing any lingering hope for a surprise defeat of Nixon at the polls.

The only surprise on November 7, 1972, was the magnitude of Richard Nixon's victory over George McGovern. Forty-nine out of fifty states voted Nixon into a second term in a landslide of massive proportions. The American voters had yet to see the darkest side of Richard Nixon.

When Thieu wouldn't budge, the negotiations between Kissinger and Le Duc Tho bogged down. Nixon decided to resume bombing North Vietnam. On December 18, B-52s began nearly two weeks of the most intensive bombing of the war.

Carol, Anthony, and I went to Texas for Christmas that year. About all that I honestly remember of that trip was listening night after night to the news of the awful bombing campaign that Nixon and Kissinger waged right through the holiday season. Nearly forty thousand tons of bombs were dropped, mainly in the populated sixty-mile stretch between Haiphong and Hanoi, between December 18 and December 30. Mainly strategic targets were hit, but civilians suffered casualties as well, including eighteen killed in the bombing of the Bach Mai Hospital. United States losses were significant. Fifteen American B-52s were shot down—the most in the history of the aircraft—accounting for more than 10 percent of the entire Southeast Asian fleet. Each one of the lost planes cost more than it would have taken to save the College of Pharmaceutical Sciences.

Finally, the worst of everything was over. Hanoi signaled its willingness to resume negotiations if the bombing would stop, which it had by the time we returned to New York. And in my mail were letters from two scientific journals, accepting the two papers I had resubmitted. Within days, Jerry had informed me that the Department of Physiology had shown some interest in me for a position and might be contacting me soon.

In Saigon, Nguyen Van Thieu realized that the United States was ultimately going to make peace with or without him, so negotiations moved rapidly toward an agreement in the second week of January. The treaty was finally signed, bringing to an end the nation's longest (at that time), most inconclusive war at a cost of more than forty-five thousand American lives. The cease-fire became effective at 7:00 p.m., Eastern Standard Time, on January 27, 1973. A cold, gentle rain was falling on New York City that evening, cloaking the skyline in a somber, muted haze. The provisions of the treaty were essentially the same that had been agreed to in October, and no reasonable person expected the "enduring peace" that Nixon was proclaiming. Whether he and Kissinger themselves really believed it, I don't know. Here and there, a church bell rang, but there was no rejoicing in the streets, no elation, no sense of fulfillment—only a deep sad sigh of relief.

Lyndon Johnson did not live to see the end of the war that became his downfall. He died of an apparent heart attack at his Texas ranch on January 22, five days before the cease-fire. I was stunned and saddened to a degree that surprised me. Ours had been a tempestuous relationship. I remembered with clarity the November morning in 1964 when I had walked through a sandstorm in Lubbock to vote for the first time, casting my ballot for a fellow Texan, Lyndon Johnson, for President. I never forgot the admiration I felt and the gratitude that the nation felt over the forceful way that he held the country together psychologically in the aftermath of Kennedy's assassination. For a brief period, as a nation, we managed to set aside our tribalism and our internal cold-war antagonisms. We enabled the most progressive social legislation of the century, including monumental advances in civil rights that only the southern conservative Johnson, not the northern liberal Kennedy, was able to get through Congress. That so much of Johnson's Great Society floundered in subsequent years was testimony only to the fragmentation of good intentions that war-drained resources and political dissidence brought about, rather than a lack of ambition. My parting with Johnson over the issue of the war by the end of 1966 was a sad breach in our relationship. Even there, his intentions had been noble, if his reading of history had been flawed; and by 1968, he had seemed on the threshold of realizing what it would take Nixon four more years and fifteen thousand American lives to see though never admit: that the outcome of the civil war in Vietnam was beyond our control. Johnson's regional defensiveness and streak of anti-intellectualism—which, being from the same part of the country, I understood—were crippling to him; yet his vision and compassion made him a giant among the small-minded men that succeeded him. In coming to grips with my feelings about Lyndon Johnson's death, I decided to try to remember him for his idea of what a great society could be and should have been, rather than what it turned out to be.

Summoned to Detroit

Eight days after the death of Lyndon Johnson, and three days after the official end of United States involvement in Vietnam, I received a letter from Walter Seegers, Chairman of the Department of Physiology at the Wayne State University School of Medicine, inviting me to come to Detroit to interview for the position of assistant professor in his department. I flew there on February 25, 1973, for my two-day interview at Wayne State.

Walter Seegers and I hit it off well, no doubt in part because I had taken the hint to read his book *My Individual Science* in some detail before my arrival. Within minutes, we were into what was obviously his favorite subject: the bizarre deconstructionist philosophy that he had adopted late in life. It was Seegers' view not only that objective reality does not exist but that reality itself is an illusion. Needless to say, this ran contrary to the fundamental assumptions of mainstream scientific thinking. Seegers was not the least concerned that the vast majority of people, including nearly every scientist on the planet, disagreed with his fundamental premise; and his tilt toward unconventionality made him tolerant, for instance, of a person like Georges Ungar, whose recommendation I was counting on to help get me the job if it was offered to me. All this came out in a lively conversation within the first hour of my meeting with him. I think the very fact that I showed any interest in what others had routinely dismissed as the foolish musings of an aging scientist made a strong early impression to my benefit.

Robin Barraco was the first faculty member after Seegers that I met. If the chemistry between Seegers and me had been good, the affinity between Barraco and me was electric and instantaneous. Incredibly articulate and enthusiastic, Barraco believed that uncovering the mechanisms of memory lay just around the corner of a little hard work and a few ingenious experiments. His own paradigm was operant conditioning in pigeons, and he had set up an elaborate lab for automating their learning behavior. In combination with my neurochemical approaches, it occurred to us both right away that together we conceivably could make a very powerful team.

By the time I had finished my first rounds with Seegers and Barraco, my head was swimming, and it wasn't even lunchtime. The rest of the day, perhaps fortunately, was less spectacular, culminating in my seminar presentation on "Specificity and Plasticity in the Brain: Problems and Approaches," which seemed to be well received. Over lunch the following day, Seegers confirmed that he was prepared to offer me the position. We briefly negotiated, if you want to call it that, over salary. Having worked at the frozen rate of $12,000 a year for nearly three years, I meekly indicated that I hoped my teaching experience beyond the postdoctoral would be worth something in the $14,000 to $15,000 range. Seegers replied, "We're going to do better than that," and offered me $16,600, which happily ended the conversation. I flew back to New York on the last day of February, knowing that I had survived for a second chance at a career in neuroscience. As the finger lakes of upstate New York slipped westward, the sun was shining brightly on a new day in more ways than one.

Saying Goodbye on a Cold, Rainy Day

The semester ended and, with it, my job and our lives in New York. I was not sorry to leave the College of Pharmaceutical Sciences. Though sentimentally attached to the building that had housed my first office, lab and classroom as a faculty member, the institution was clearly an anachronism whose demise had been hastened by poor administration. It probably didn't deserve to survive. But the students were another matter. In my three years there, I had seen some great young people come to study pharmacy, had felt a kinship with their struggle to be serious students during the chaotic upheavals of the early seventies, and had grown close enough to a few of them to count them as personal friends: Bonnie Harting, Sam Liss, Franz Ozalas, Bruce Gray, Mohammad Zaman, Dennis Kocjan, Donna Hills, and Judy Mancini.

One of the final acts of the school year was an outing that I took with a number of students to Shea Stadium for a Sunday afternoon of baseball. The weather conspired against us, however; the day turned cold and gray, and rain began to fall before game time. We waited in vain as the temperature dropped and the rain intensified. Finally, the postponement announcement came, and we rose in dismay to trek to the subway for the hour ride back to Manhattan.

Judy Mancini was standing a few seats from me. We looked at each other, knowing it was probably the last time we would see each other again. Of all the students I had come to know in New York, none had touched me more than Judy. No one had been more devoted to our shared attempt to do research under trying circumstances. Our relationship had grown strong and deep and mutually reliant. I wanted to let her know how much inspiration she had been to me in those first tenuous years of my career and tell her that I thought the world of her and give her a long lingering hug of goodbye. But both of us were too reserved for that, and besides, we were in the middle of a huge crowd. All we managed to do was mumble a muffled and sad goodbye to each another as we headed for the exit.

The rain continued for a couple of days afterward, and I couldn't shake the depression of that final scene. Finally, to clear it from my consciousness, I wrote a poem starting and ending with the same two lines:

> We say goodbye on a cold, rainy day
> In the middle of a thousand people.

The time for our departure from New York drew near. Carol and I saw much of the city in those final days, drinking in as much as we could, as exiles must prior to their banishment from a treasured place. We purchased a used Volkswagen camper and loaded it with our laundry and personal belongings while a moving van packed our books, furniture, and household belongings, including the cockroach eggs that would hatch to populate our new home in Detroit, into a truck for the trip westward.

The final evening before our departure, I took a last walk to Central Park. It was a lovely summer afternoon. The *Playboy* and *Penthouse* softball teams were locked in combat while thousands of others on bikes or on foot, with dogs, baby carriages, Frisbees, or brown paper bags, rode or strolled or played or staggered through the dwindling sunlight of the waning day. I thought of this richest character of the city—its teeming and varied populace—of the surrounding skyscrapers from which they had poured a few hours earlier into this magnificent and expansive central playground; of the stores and shops that made anything available within walking distance; of the subway system that, for all its noise and graffiti, had whisked us wherever we wanted to go in a matter of minutes, rendering autos an encumbrance; of Lincoln Center, which had brought me the excitement of opera and ballet for the first time and the live symphonic music and theater that I relished, just two blocks away. I had come to care very much for this place and had not yet had my fill of it. I breathed in the air of Central Park for one last time, freezing as long as I could the scent, the sound, and the view of that place and time in my memory. Life was moving along, and I knew that another time and place awaited me.

Carol, Anthony, and I crossed the George Washington Bridge the next day at twelve thirty in the afternoon traveling west, only three and a half short years after the September morning in 1970 when I had crossed it in the other direction with so much hope and expectation.

[1] This section on the search for and discovery of the opiate receptor is based mainly on the excellent book by Jeff Goldberg: Goldberg, J. 1988. *Anatomy of a Scientific Discovery: The Race to Discover the Secret of Human Pain and Pleasure.* New York: Bantam Books.

[2] Goldstein, A., L. I. Lowney, and B. K. Pal. 1971. Stereospecific and nonspecific interactions of the morphine congener levorphanol in subcellular fractions of mouse brain. *Proc Natl Acad Sci USA* 68: 1742–7.

[3] Terenius, L. 1973. Stereospecific interaction between narcotic analgesics and a synaptic plasma membrane fraction of rat cerebral cortex. *Acta Pharmacol Toxicol (Copenhagen)* 32: 317–320.

[4] Pert, C. B. and S. H. Snyder. 1973. Opiate receptor: demonstration in nervous tissue. *Science* 179: 1011–4.

[5] Simon, E. J., J. M. Hiller, and I. Edelman. 1973. Stereospecific binding of the potent narcotic analgesic (3H) Etorphine to rat-brain homogenate. *Proc Natl Acad Sci USA* 70: 1947–9.

MIDSEVENTIES (DETROIT)

22

The Gift of Touch

I finally started coming to grips with scientific reality during my three-and-one-half years of living in Detroit. I did so partly because of, and partly in spite of, the most overpowering and in some ways most impressive person I ever came to know.

Robin Anthony Barraco

During my interview at Wayne State in February 1973, Robin Barraco and I hit it off well. In deciding to accept the job in Detroit, one of its major attractions was the prospect of collaborating with him. Our limited correspondence during the spring prior to my arrival reinforced all those good first impressions. Even so, my first few weeks in Detroit in the presence of Robin Barraco brought a level of stimulation and excitement to my intellectual life that surprised me and exceeded all my expectations.

Robin was one of those people who made an overwhelming first impression. His energy level and effusiveness were dazzling. He cut through formalisms immediately, impressing with energy, disarming with earnestness, flattering with courtesy and praise. Supremely confident at all times, his quick-witted intelligence powered an articulate stream of conversation that compelled his listeners to the belief that he must have known everything of importance about anything. Yet in his presence, if he wanted to, he made you feel almost as smart and perhaps more worthy than he was in spite of seeming evidence to the contrary. He had the ability to reach into your being, touch your heart, and wrap himself around your mind (or, if you weren't careful, simply seize your brain). Those who collided with him, for a day or a decade, knew they had been hit—whether by substance or illusion became a matter of retrospective analysis, but they knew they had been hit by something out of the ordinary.

Robin Anthony Barraco was born on March 24, 1945, in Detroit of Roman parents and Sicilian grandparents. Often during the time I knew him, if he felt the need to grimly impress or frankly threaten, he would remind his listener, "My family is Sicilian." The eldest of three children, he grew up in Detroit and was graduated from the University of Detroit Academy in the top five percent of his class. He was accepted to the study of history and philosophy at Georgetown University in 1962, his aim being to become a college professor of philosophy. He was particularly interested in the theory of knowledge and the structure of consciousness. In the course of devouring this subject, he came across an article by Linus Pauling on orthomolecular psychiatry. This, of course, raised the specter of a chemical basis for consciousness and led Robin to wonder if consciousness has "an epistructure that we don't understand . . . that is chemically based."

The time of tumult on college campuses was beginning, and Robin crossed easily into the counterculture of the era—reading Timothy Leary, perceiving the science of psychopharmacology converging with the philosophical lifestyle of the acid heads around him. The power of physics began to impress him as much as the probity of philosophy, and he started to think of science as the best route to genuine insight into the ultimate nature of consciousness and free will.

Driven by a libido that matched his philosophical intensity, he sought the company of many women and enjoyed the pleasures they brought him with frequency. But in time, he desired the longer-lasting comfort and constancy of a relationship with one woman, marrying a girl from Detroit in 1965. His wife wanted to return to their home, so they moved back to the Motor City after Robin completed one year of postgraduate study at Georgetown. The marriage didn't last, but by the time it ended, he had entered graduate school with the endorsement of Ray Henry in the Department of Physiology at the Wayne State University School of Medicine in Detroit. From 1967 to 1969, he held a National Science Foundation Fellowship. His brilliance was evident to all, but his lifestyle and appearance were too eccentric for Seegers, a courtly gentleman of conservative tastes. Robin passed his first doctoral prelim—an examination on the work of Pavlov—with distinction, but Seegers refused to certify Robin's candidacy for the doctorate unless he shaved his beard and cut off his ponytail. Robin refused, and his progress was put on hold. Sensitive to the eccentricities of both his senior colleague and his junior understudy, Ray Henry mediated the estrangement with patience and eventually brought Robin back into the fold.

In 1969, Robin was struck with hepatitis—an experience that characteristically launched a reading foray into the literature on antibiotics, then antibiotics and memory. The science of mind according to Agranoff, Barondes, and Flexner was opened to him, and he resolved to master the intricacies of behavioral biochemistry. Still as a graduate student, he began to teach segments on the biochemistry of behavior in undergraduate psychology courses.

Upon completion of his dissertation, so complete was his rehabilitation that the clean-shaven Dr. Barraco was hired by Seegers as an assistant professor of physiology in the department, which lacked at that time anyone with expertise in the nervous system. Seegers and Barraco, perhaps finding common ground in their mutual passion for philosophy, drew curiously close; and Robin was one of the faculty members entrusted with reviewing resumes for an opening for another assistant professor. It was at this juncture in both our careers that Robin Barraco became aware of my existence.

There were several factors that contributed to the powerful and almost immediate chemistry between Robin and me. Though clearly a young scientist of extraordinary intelligence, he had not trained in the lab of any well-known neuroscientist and didn't even know any scientists personally outside of Detroit. Since I was very familiar with the conventional side of neuroscience and was acquainted with many of its prominent practitioners, I provided an introduction for Robin to the conventional scientific community.

We both shared an unbridled enthusiasm for the work of science— the sheer pleasure of laboring in the lab, working with animals, manipulating technology, generating data, and playing the mind games that tried to make sense of it all. Most importantly, we shared the belief that the mechanisms of memory are knowable at the molecular level and that, with hard work, we had as good a chance as anyone of figuring out what they are.

Beyond our affinity as scientists, however, Robin reached me as a person in a way that no one else ever had. I was German, English, and Scots-Irish by family history, with all the Calvinistic reserve implicit in that ethnic background. Though laced with the southern veneer of a congenial personality, I was fundamentally introverted and dispassionate except in very private settings. Robin was Sicilian (he reminded me often), with the power and volatility of Mediterranean emotions driving all his actions. To his hurry-up northern mentality was added a level of energy and ability that gave him the personality of a jet-powered bulldozer. He was sensitive to the differences between us and respectful of them. But his experience with people of my temperament had been that they could be salvaged by the appropriate amount of good-humored goading, so he began to work on me almost immediately—stirring the animation and spontaneity of which I was capable but not inclined to let out very often, reinforcing a more open display of my state of mind, and encouraging me to loosen up and get in touch with myself and others better. From as far back as the painful days of Houston, I had been aware of the other side of my nature—the impulse to dance on the beach, like Zorba, or flaunt my colors, like Braniff's brightly painted jetliners. Robin touched that part of my inner being and helped me begin to give vent to it, ever so slightly and in extreme moderation. He forced me out of my comfortable demeanor often enough to begin to change just a little bit not only the way I behaved, but the way I looked at the world. Among other things, he taught me how to touch other people, literally, as a show of human connection. He was forever hugging me or kissing me on the cheek (which he so enjoyed because he knew how much it mortified me) or patting me on the arm just in saying hello or goodbye. After I had picked up on this and realized that the invisible barrier between people could be breached effectively by a quick, gentle touch appropriate to the moment or a brief hug among friends, I felt that my capacity for positive human interactions had been vastly improved. When, in later years, I tried to assess the mixed impact that Robin had had on my life and career, I realized that the one impact that stood out, and the influence that had benefited me the most, was an enhanced capacity for the physical expression of affection. Robin gave me excitement, praise, and friendship; but most of all, he gave me the simple gift of touch.

Resurrected Research at Wayne State

The second day I was in Detroit, Robin held an all-day research meeting with Larry Stettner and their team of graduate student and undergraduate researchers. I was invited to participate, to say a little about my background and research interest, and to contribute whatever I liked to the discussion. It was the first time since my postdoctoral year in Kansas that I had sat down to talk about the details of actual experiments with colleagues of like-minded enthusiasm. I was welcomed into the group immediately and made to feel that I would be a vital part of the team.

To get us started with something that might be achievable and publishable quickly, Robin and I set out to (1) test for the presence of detectable peptides in the pigeon brain and (2) continue the analysis of sialidase activity in rat brain tissue that I had begun in New York. The specter of scotophobin still lurked in the back of my consciousness, and the recent discovery of endogenous opiate receptors was rapidly reviving interest in the possibility of neuroactive peptides. Gene Murano, another junior faculty member in the Department of Physiology, had worked out a technique for separating small amounts of peptides from other compounds by two-dimensional electrophoresis, so Robin and I saw the possibility of testing for the presence of endogenous peptides in the pigeon brains that we were collecting as part of our behavioral studies.

By shortly before Christmas, we had clearly demonstrated the presence of half a dozen or more peptides endogenously present in extracts of pigeon brain. Seeing the first chromatograph with the telltale purple spots giving positive evidence of peptides was a dramatic moment of great excitement for Robin and me. Our first mutual discovery, it reinforced our belief that our teamwork was going to bring us great success.

Geopolitical Disasters Here and There

The drama of that early breakthrough has merged through the years of my memory with a number of other dramatic events that drew 1973 to a close. In September, a military junta had overthrown the government of Salvador Allende in Chile, the first democratically elected socialist head of state in the Western Hemisphere and the first violent overthrow of government in Chile in half a century. It later came out that the CIA had played no small part in fomenting the political unrest and economic instability that gave the military its excuse for instigating what would become years of a ruthless dictatorship. The troubled Middle East exploded into war once again on the holy Jewish holiday of Yom Kippur in October, with a coordinated invasion by Egyptian and Syrian forces into the territories occupied by Israel. By the time a cease-fire had been agreed to sixteen days later, the Arab states were worse off militarily; but Egypt, at least, had regained the self-respect needed for making the overtures of peace toward Israel that would come a few years later.

Overriding all these events in the national news were the continuing revelations of crimes and misdemeanors at the level of the presidency related to the Watergate affair. When they were eventually made public, they doomed the presidency of Richard Nixon.

In the midst of all this, Vice President Spiro T. Agnew, who had raised rhetoric and vilification of political opponents to new heights and had made law and order his crusading theme, resigned the second highest office in the land on October 10, pleading no contest to federal charges of income tax evasion. As it turned out, the law-and-order crusader was himself apparently a major violator of the law.

So as 1973 came to a close, the national and international political scene was in disarray. But for me professionally, the year had been a bright and hopeful turning point. Carol was fortunate as well upon arriving in Detroit to find a postdoctoral position in the Biochemistry Department at the medical school that was compatible with her interests and background. So by year's end, our professional and personal lives were on a comfortable footing for the first time in our marriage, and we looked forward to building on that stability in the year to come.

23

Metabolic Profiles

As 1974 got underway, Robin and I were plunging forward with a comprehensive analysis of macromolecular turnover in the pigeon brain during operant conditioning—what we referred to informally as the Bogoch experiment because it was an elaboration of Samuel Bogoch's analysis of glycoproteins from trained pigeon brains, but which we gradually began to call our metabolic-profile approach in more formal and polite company.

It was frankly a surprise when our early data came out suggesting that Bogoch had been right to a degree. Like that of Bogoch, our fractionation procedure divided the glycoproteins according to their degree of acidity. Some of the more acidic glycoprotein fractions from trained pigeons accumulated more radioactive labels than the same fractions from control pigeons. One or two other fractions claimed by Bogoch to show learning-related changes did not do so in our experiment. If anything, this disparity bolstered our confidence in the reliability of our data; we felt our experiments were rigorous and reproducible enough to stand up to the skeptical scrutiny that we expected. Given Bogoch's more flamboyant style and extravagant interpretations, it was frankly a relief not to find exactly what he claimed. We basked in another small but significant success and made plans to present the data at the New Orleans meeting of the American Society for Neurochemistry.

Overdue Recognition and a Cool Reception in New Orleans

In New Orleans, the presentation of our results was received by a limited audience with a little interest but not much enthusiasm. As I expected, there was skepticism over this relatively crude, shotgun approach to searching for molecules that might have something to do with memory, among a vast array of metabolites that surely had nothing to do with memory. But there was at least one enthusiastic observer; Samuel Bogoch, not surprisingly, was quite impressed—to the point of dragging others over to our poster to point out the elegance of our work. Whether this acclaim from Bogoch did us more good than harm was debatable.

On a personal note, I finally received the professional recognition of election to full membership in the American Society for Neurochemistry at this meeting. Even though I was a faculty member and independent investigator by the time of the second annual meeting of the Society in 1971, for unaccountable reasons, a succession of membership committees had declined to elevate me beyond essentially student status. One officer in the society, Marjorie Lees, had argued in favor of my promotion earlier, but she remained in the minority. I was paranoid enough to suppose that my work in controversial areas of neurochemistry and without the backing of an established, mainstream neurochemist, had left me a little suspect. Finally, with half a dozen publications as sole or senior author, the merit of my case had become undeniable. I remained a little bitter over this for many years.

I returned to Detroit from New Orleans via Evansville to visit with Bob and Muriel Godbout. It was a wonderful and all too brief reunion with these closest of friends, and I wouldn't have traded it for what Robin endured after the meeting for anything. But what Robin endured was admittedly more exotic. Characteristically restless, he had spontaneously decided on a foray into Mexico for a few days of color and excitement before returning to the drabness of Detroit. In Merida, he met up with a couple of Western Airlines flight attendants who invited him to join them on their tour of the Yucatan Peninsula. For the next few days, Robin charmed and delighted his two traveling companions—one, thirty-two; the other, forty (Robin was twenty-eight at the time)—sparing none of himself to satisfy their considerable sexual appetites, though taking care to consort with them one at a time, making each feel special as he tended to do with all the women he knew. At the end of their adventure, he left them satisfied and apparently wanting more, for a few months later, they sent him a ticket to the West Coast that he sadly had to decline under the pressure of other obligations.

Daylight Develops between Robin and Me

The metabolic profile strategy that Robin and I had now committed to was very labor intensive. For statistical reasons, we had to use many animals, and there were different brain regions and many chemical fractions from each region to be processed. The behavioral work itself took time, and the chemical extractions and analyses took a lot more time. As junior faculty members with considerable teaching duties and no grant support to hire technicians, I felt it essential that we stay focused on the essence of the experiment, keeping the methodology as quick and simple as possible. It was Robin's nature, on the other hand, to burrow into methodology in exquisite detail, to salvage and process every conceivable fraction, and to follow up on every intuitive lead. During the early months of our collaboration, we compromised these disparate inclinations, but it became clear that a resolution of our different approaches would be needed down the road.

It was another measure of the difference between Robin and me that I found writing papers and grant proposals less onerous than did Robin, so he was happy to leave the writing to me while he busied himself with minutia in the lab and the high-energy back-slapping camaraderie among our colleagues that was important from a public relations point of view, especially given the queasiness that some of our colleagues felt about the whole field of behavioral neurochemistry.

Our differing lifestyles in time became a factor as well. A bachelor with an active social life, he liked to work at odd and unpredictable hours. I had a family, and though I neglected them badly, my schedule was constrained by babysitters' schedules and the good faith intention to spend a few predictable hours a day with my wife and son. I could tell that it drove Robin up the wall when I would have to stop an assay or extraction at five o'clock in order to pick up Anthony or ride home with Carol. To his credit, though, he respected my personal choices and understood the nature of family life. At some level, I suspect he may even have envied my domestic tranquility, by contrast with his hectic and volatile personal relationships.

By June, these various differences were beginning to intrude into our working relationship. I believe that the cool reception to our paper and the criticism of our methodology that we experienced in New Orleans was a bit of a shock to Robin and reinforced his concern that our methodological approaches were not sufficiently comprehensive as he would have liked. But both of us had gained so much from our collaboration, had—in my case—literally experienced a professional resurrection, that splitting up was inconceivable. We chose instead to view our differences as complementary strengths. But noting our areas of incompatibility, we decided that the better tack for us to take would be to develop different, though related, projects all under the same conceptual umbrella, but each totally under the control and execution of one or the other of us. He would concentrate on the sophisticated biochemical fractionation techniques, pursuing problems in-depth at the chemical level. I would work on metabolic profiles and other projects that depended on broad sampling across many anatomical, subcellular, or gross molecular populations. That way, we could enjoy the benefits of our collaboration without having the idiosyncrasies of either of us interfere with the other. It was a good decision, and it stabilized our relationship for the remainder of my years at Wayne State.

With that issue out of the way, I plunged into my writing assignments. While it was true that I handled the writing portion of science with less discomfort than Robin, it wasn't my favorite part of the work. I enjoyed the actual labor of the lab to a greater degree: working with my hands to set up a complicated apparatus or do small animal surgery, watching spots come into view on a chromatogram, collecting liquid fractions from a separation column and measuring the color intensity of a biochemical assay, realizing that only a handful of people in the world could draw a logical connection between the color of a solution in a test tube and a biological process in the brain of a pigeon or, someday maybe, a human. These day-to-day aspects of the rhythms of research were the side of science I most enjoyed, and diversions from them for any length of time inevitably led to a growing frustration. So the summer wore on productively, but not as enjoyably as I would have liked.

Meanwhile, on the Political Front

There was another diversion, however, that I found to be rather exciting. Always politically aware, my interest in politics was urged to a new level by a brief editorial that appeared in the *Detroit Free Press* in early May. It described the precinct delegate system in Michigan, whereby representatives from each precinct were elected in August to serve as delegates to the state conventions of each party. Only fifteen signatures were needed to get on the ballot. The editorial pointed out the ease and utility of using this mechanism to get into politics at the most local of all levels. I couldn't resist. Armed with a clipboard and pen, I took the plunge that all unknown but aspiring politicians have to take sooner or later: I began to trek from door to door in my neighborhood, introducing myself to total strangers, asking for their vote for me for an office that most of them had never heard of. I was buoyed and somewhat surprised at the receptiveness of my neighbors, most of whom were African Americans and had reason enough to be leery of white strangers. But that type of tolerance was the rule rather than the exception in the integrated Boston-Edison neighborhood of inner-city Detroit. Getting the fifteen names wasn't hard, and I found that I enjoyed the subsequent door-to-door campaigning more than I expected. When the votes were in, twenty-eight had voted for me and twenty-six for my opponent. I was probably the first white precinct delegate from the area in at least a decade.

As my political career was beginning, Richard Nixon's was ending. On August 5, apparently under pressure from his attorney and senior staff, Nixon released the transcript of his conversation with H. R. Haldeman a few days after the Watergate break-in, which clearly confirmed their conspiracy to cover up the true facts of the incident by portraying it as a CIA operation. Reaction in Congress and throughout the nation was immediate and vigorous. Faced with the certainty of impeachment in the House and conviction in the Senate, Richard Nixon announced Thursday night, August 8, that he would resign at noon the following day.

I really took no pleasure in the demise of Richard Nixon, other than the self-satisfaction of being among the minority of Americans who had said in 1968 that electing him would be a mistake and, in 1972, that reelecting him would be a tragedy. Had he shown more compassion for the thousands of Vietnam War resistors that he turned into fugitives or to the tens of thousands of allied soldiers and Vietnamese soldiers and civilians that died as he prolonged the war for four unnecessary years for "peace with honor," it would have been easier to summon some sympathy for this victim of the greatest political fall in American history. But there was certainly no rejoicing by me or by anyone I knew over his sad, pathetic demise—only a melancholy sense that it had somehow been inevitable. Having often appealed to the worst in others, Nixon did not realize that most people could be charitable and tolerant of mistakes. He thought, instead, that he was greater than he was, that he knew more than he did, that he could get away with more than he could, and that he was besieged more than he was. It was his own weaknesses and limitations that were his downfall. In that sense, he deserved his fate, and we were better off without him.

Partly because of the political turmoil of the summer and largely because of the short attention span and ephemeral focus of the American character, the fifth anniversary of the first human landing on the moon went barely noticed. Two days after the date, a short story appeared on page 12 of the third section of the *Detroit Free Press*, noting,

> Neil Armstrong, the man who did it, called it a "giant leap for mankind."

President Nixon called it "the greatest week in the history of the world since creation."

The American people called it too expensive.
Within a year after the flight of Apollo 11, nearly two-thirds of the people surveyed in a Louis Harris poll said space exploration wasn't worth the money. The Harris organization has never since felt any need to ask that question.

The article went on to quote a statement by former Apollo 11 operations director John K. Holcomb that summed up my feelings exactly:

Look back in the 1960s at what this country did that was universally respected and admired, and it's hard to think of anything else.

With the possible exception of striking advances in molecular biology and neuroscience, I might have added.

24

Disillusioned and Despairing in St. Louis

The city of St. Louis is not generally regarded as a major scientific center, but Washington University in that city has played a special role in the advancement of neuroscience. It was here in the 1920s that Joseph Erlanger, Herbert Gasser, and George Bishop first put together the electronic equipment capable of measuring the tiny electrical activity of the nerve impulse. Francis O. Schmitt, one of Erlanger's early students, would become a major figure in the historical development of both biophysics and neuroscience, founding the Neurosciences Research Program in 1962. But in particular, St. Louis and Washington University have had a special impact in the field of developmental biology. Victor Hamburger, master of embryonic microsurgery and author of many conceptual and technical advances in the study of development, succeeded Frank Schmitt as chairman of zoology at Washington University. Rita Levi-Montalcini did some of the critical work on nerve growth factor, a molecule of major importance in spurring the outgrowth of neural processes, in Hamburger's department. Maxwell Cowan and his students illuminated the details of neural development in both the chick retinotectal system and various mammalian neural structures as no other lab has done, and to this day, an outstanding cadre of younger neurobiologists continues to set the standard for research on the development of the nervous system at Washington University.

Anticipating a Triumph

For well over a year, I had known that the annual meeting of the Society for Neuroscience in 1974 would be in St. Louis that October. It was natural, given the importance of St. Louis as a center for research in developmental neurobiology, that I would think of that meeting as the obvious platform for the first presentation of my developmental work. Hamburger and Cowan themselves, and maybe even Roger Sperry—the originator of the chemoaffinity hypothesis—would presumably be in attendance. Thus as the summer ended, I turned my full attention (as much as could be spared from my growing teaching duties) to the analysis of biochemical changes in the developing tectum of the chick embryo. If I could show that the ratio of sialic acid to carbohydrate, or of carbohydrate to protein, differed at different points in the optic tectum, or that different gangliosides marked different positions in the tectum, then a chemical basis for site-specific coding of cells as demanded by the chemoaffinity hypothesis would be demonstrated. Drawing on the microdissection techniques learned from Byron Wenger, one of my professors at Kansas and a student of Hamburger, I divided the tectum into tiny topologically distinct regions and proceeded to carry out the appropriate biochemical analyses on each region separately.

To get enough tissue to analyze required surgery on many dozens of embryos, so again I was embarked on a very labor-intensive project. But the effort paid off as the data came in. To my delight, there were indeed regionally specific differences in both the sialic acid–carbohydrate ratios and in the carbohydrate-protein ratios. To my surprise, however, these differences were not static, but oscillated over the time during which retinal fibers were growing into the tectum. So the retinal fibers indeed had chemically distinct targets to attach to, but the targets appeared to be moving in time. Successful neural connections between the retina and the tectum might require coordination in both time and space—a fact that the morphological evidence by then was suggesting but which had never been demonstrated at the chemical level. The evidence for regional differences in ganglioside distribution within the tectum was clear as well. Gangliosides, then, might also be providing molecular signposts for directing the appropriate neural connections. This was even more exciting because the precise chemical structure of the gangliosides, and much about the enzymes that controlled their synthesis and degradation, were known. So it was at least conceivable that a link could be drawn between the genetic information of the cell (which codes precisely the enzymes that control the production of gangliosides) and the specific organization of the brain, and all that this implies for mechanisms of information processing. As the time for the St. Louis meeting approached, I felt myself quite possibly on the verge of joining the corps of recognized leaders at the forefront of research on neurospecificity and was pleased at the thought that they would all be there to hear and see what I had done.

The session into which my presentation had been scheduled could not have been more appropriate. All the other papers were on some anatomical aspect of the retinotectal system. Mine was the only presentation to deal with the chemical aspects of retinotectal development, and it was the last paper of the session, which was perfect since the papers to precede it were excellent setups. I did note ominously that the time for my presentation was 4:30 p.m., the prime cocktail hour on the fourth day of a heavy meeting. But given the power of my message, I wasn't overly concerned.

The session lived up to its advanced billing. There were papers by Maxwell Cowan, Sansar Sharma, Myong Yoon, and Carla Schatz—all established or emerging figures in developmental neuroscience and subscribers, in varying degrees, to the chemoaffinity explanation of the formation of specific neural connections. Several hundred people attended, some of them vigorously debating the data and conclusions offered by authors from the podium. One of the more dramatic presentations was from a young woman at Cornell, Martha Constantine-Paton, who had induced the formation of a third eye in a frog and studied the degree to which nerve fibers from this extra eye could compete with fibers from the two normal eyes. There was tremendous interest in her presentation, which was delivered with verve and apparent self-confidence. Clearly, she was a star in the process of being born.

Talking to a Sea of Empty Chairs

This was the seventh paper I had given at a national meeting, and none save the first had been prepared with more thought, care, and rehearsal. As the speaker just prior to me was ending, I gathered my nerve and energy for the presentation of my career. But before my name could be called, to my astonishment, then sickening dismay, an exodus from the lecture hall of massive proportions ensued. Well over a hundred listeners had stayed till four fifteen to hear a talk on decreased cerebellar DNA in malnourished rats, but in the following dreadful moments, it became obvious that almost every one of them had heard all the science they cared to on that afternoon. Who wouldn't be weary after more than three hours of developmental neuroscience? I understood that. But for another fifteen minutes, they could have heard the first breakthrough on the chemical side of the equation. That they didn't really care was incomprehensible.

I stepped to the podium in a daze, then forced myself to look out across the empty sea of chairs. For a brief moment, my mind leapt to the conclusion that no one remained at all, that I could skip the presentation of my paper and, by some twist of poetic justice, simply vaporize to nothingness as my audience had done. But as my eyes focused on the dim, empty vastness, I counted—twice to make sure—eight bodies still stuck inexplicably to their chairs. One of them was a former student, Bruce Gray, and three others were friends of his or mine from Wayne State. That left four strangers to hear what I had to say; and I say strangers advisedly because I would have recognized Hamburger, Cowan, or Sperry.

The thought of apologizing to the hapless chairman of my session and dismissing the loyal audience of eight without giving my presentation did seriously occur to me; but an instinct borne, I suppose, of my days in high school drama told me that the show had to go on. I wasn't the first scholar in history to give a paper to a small, tired audience during the cocktail hour, after all. Once into the talk, my momentum carried me; and by well into the talk, I was able to emote some enthusiasm, such was the beauty of the data and the importance of the findings I still believed. The spatiotemporal oscillations in chemical characteristics of tectal cell membranes were real, whether anyone was there to hear about them or not. And gangliosides were distributed over the tectum asymmetrically, whether anyone thought it important or not. It turned out to be one of my best presentations ever.

That night, I dined alone. In the privacy of my thoughts, I turned over the events of the day and, gradually, very gradually, began to appreciate the extent of my delusions. It wasn't that the work had been poor or that it lacked significance. It was that nobody cared. What matters for scientific acceptance is not how good or important something is, but how relevant it is to what other scientists care about. And that usually translates into what they are doing. Nobody was working on the chemical side of chemoaffinity because they were too busy conducting ever more elaborate anatomical manipulations (growing three-eyed frogs, for example) or because they didn't know how to take a credible chemical approach. And if they weren't working on it, they didn't care about an isolated researcher whose work bore no chance of supporting or refuting their own.

But if this were true, what was the point of doing anything out of the mainstream? And for certain, what was the point of killing myself to get abstracts in by a deadline, conduct experiments to the last minute, and go through the agony of countless rehearsals for a presentation that interested almost no one? I couldn't answer the questions that night, for I knew I couldn't think about them with any objectivity at all. I went to bed and finally to sleep, feeling bitter, proud, and sad.

In my wildest dream that night, I could not have imagined that about a decade later, Martha Constantine-Paton would discover a molecule asymmetrically distributed in the retinotectal system of certain mammalian species and that, with the help of some talented biochemists, she would come to discover that it was a ganglioside, further enhancing her already impressive reputation for trendsetting research in neuroscience. When, a dozen or so years after the St. Louis meeting, I walked up to her beside the poster that described her latest work on this ganglioside to ask her some questions about how it might operate to establish specific connections, she was first evasive, then clearly confused by my question. And she obviously had no idea who I was.

"I Thought I Might Find You Here"

Dianna Redburn finished her doctoral dissertation under Fred Samson's direction in 1972 and moved to a postdoctoral position in the lab of Carl Cotman at the University of California in Irvine. There she encountered an energetic, thriving, often contentious, but close and supportive group of young scientists busily turning the Department of Psychobiology at Irvine into a world-class center of research in neuroscience. Cotman was a tough taskmaster, the type of mentor that his students loved to hate because he could be that hard on them. But they were better for the experience, and they would go out from Irvine and spread the gospel of neuroscience to all corners of the scholarly world. Cotman could be selective in the recommendations he gave—the strength of his endorsement sometimes reflecting the position being applied for as much as the applicant—and this became a sore point when it came time for Dianna to leave what had otherwise been a very productive and enlightening two years in his lab. But he did give her a strong recommendation for a faculty position in the newly created Department of Anatomy and Neurobiology at the University of Texas Medical School in Houston, and she was one of the first faculty members to be hired by Joe Wood, the chairman, who hoped to build something in the mode of the Irvine experience in Houston from the ground up.

Dianna was attending her first Society for Neuroscience meeting after joining the University of Texas faculty, but the final day of the St. Louis meeting had arrived, and she had not yet seen her friend and lab mate from her earliest days at Kansas; so she hadn't yet had a chance to tell me of the twists in her career or find out about the turns in mine. Knowing me well, she had thumbed through the meeting program to find that an afternoon session on neural plasticity was underway in the Tiara Lounge of the Chase-Park Plaza Hotel. She headed there and found me sitting halfway back in the middle of a mostly empty auditorium (it being the last afternoon of the meeting).

She slid into the seat beside me, under cover of the room lit dimly for the talk in progress.

"I thought I might find you here."

We exchanged brief warm hellos, but talked no further out of courtesy to the speaker at the podium. When that speaker finished, she asked if I really wanted to hear the next presentation, and I said no, so we left the hall for a better place to visit.

Our small talk was brief. Suddenly, I found myself pouring my heart out to her, reliving the mortification of twenty-four hours earlier, expounding on my failures, bemoaning my fate as a reasonably intelligent scientist on the road to nowhere, attracted to areas of research that everyone found interesting but no one would touch, breaking my neck to do the right things, yet sensing, increasingly, that they weren't the right things. I pleaded for insight into the fickleness of professional science—a rationale for the collective judgment of three hundred scientists that three-eyed frogs illuminated the theory of chemoaffinity more than a little simple chemistry on tissue from embryos with just the two eyes that nature intended. Why was I spinning my wheels? Where had I gone wrong? Was there, God forbid, never much there to begin with?

Success in science, as in any competitive field, calls for a fair amount of bravado. As a man, I knew this instinctively. As a woman, she had learned to appreciate it keenly. It was thus a measure of our trust and friendship that no such posturing was required between the two of us. Whatever our later professional fates and public images, we could always return with each other to the mindset of our earliest days together when as graduate student and technician, neither of us had anything to pretend. That I could bare my soul to her in this way was thus no surprise, and it was not the first time that either of us had done so. Still, had I run into her in the coffee shop or found her strolling through the exhibits, I doubt that I would have launched into such a self-incriminating confessional quite so readily. There was something about *her* coming to *me*, seeking me out on the last day of the meeting, as though her well-being were at stake rather than mine, that had touched me and lowered my reserve. This subtle reminder of my importance to her was what had loosened my torrent of anguish.

Whatever her feelings about my choices, whatever she thought of my research, she would not, I knew, be anything but supportive; and she would mean it because, for all the more important and accomplished scientists she would come to know, I knew I held a small but special claim on her loyalty, respect, and affection as she did on mine. Yet she would not say things that weren't true, nor would she flatter my ego just to rebuild my spirit. She would say what she truly believed.

After patiently hearing me out and allowing herself to soak it in, she began a soliloquy that would help me get through my self-perceived calamity. I can't recall it word-for-word, but it went something like the following.

> Yes, science can be fickle and seemingly unfair, but you have to measure your merit in more than one way. I can't judge the importance of your work or how good it is now, but as long as I've known you, I've thought your ideas were important, and you were the one that impressed upon me the need for quality control during our earliest days in Kansas. You have many talents— keep that in mind: your patience with novices, your ability to teach, a facility for the written word. You have mattered in my life and career, and I'm sure in the lives and careers of others. If you think your research is important and your approach legitimate, you have an obligation to keep trying to convince the rest of us that they are. Whether you succeed or not, though, will measure only partly your success as a scientist and not at all your success as a person.

I knew that what she had said was true, but in that moment, it had been necessary for someone else to tell me so. Her slow, deliberate words, draped in their tinge of inflection from northern Louisiana, soothed my agitated soul and lanced the boil of self-pity that had risen over the previous night and the day now drawing toward evening. I felt better for our talk, better about myself, now able to step back and ask the same hard questions in the more objective light of another day—questions about the nature and direction of my research that the reassuring words of my close friend had not directly addressed.

I went to bed that night feeling still wounded, but better. The following day, I drove from St. Louis back toward Detroit with Bruce Gray, the student who had come with me. At a Holiday Inn near Indianapolis, we stopped for the night and, after dinner, retired to the bar where we drank beer for three hours straight and talked of many things, including the power of women to affect men's lives.

Southern Comfort

All things considered, the results of our first full year together left Robin and me disappointed. It is only with the hindsight of time that I can see how far over into the discomfort range for most neuroscientists our research strategy had moved. Our approach, both in our behavioral and our developmental work, depended on a search for patterns in a broad but superficial array of chemical endpoints. If there is anything that modem science cannot stand, it is lack of focus. Not just conceptual focus (I really think we had that), but especially technical focus. It is far better to search for a single molecule with the most current and sophisticated methods available than to pursue a population of molecules in search of patterns or designs with simpler, more comprehensive techniques.

The problem with the brain is that the information that counts for all its dynamic operations is almost certainly patterned information, and how to get to that at the chemical level has been the obstacle of the age in neuroscience. To think that Robin and I were going to do it with the tools available in the mid-1970s seems now so hopelessly naive. And there were many who thought so then, though, to be fair, we received almost without exception a respectful hearing, and that includes from review panels at NIH and the editors of various journals who, sensing our sincerity and the magnitude of our effort, bent over backward to see the merit in what we were trying to do.

Even then I sensed much of what I have stated above, but Robin and I were both imbued with the scientific ideal of pursuing an alternative vision, standing firm in the face of skepticism, meeting doubt with data. So we determined to press onward, our string of bad tidings interrupted mercifully by the holidays.

Carol, Anthony, and I flew to Houston that Christmas to begin our holidays with friends and family there. Our first full day in Houston, I called Georges Ungar, and he invited me immediately over to see him. I spent half a day in his lab, having the best visit with him and Alberte that I had ever had since working with them eight years earlier. We talked in detail about what each of us was doing. He thought my developmental work, in particular, was significant, but was less enthused about glycoproteins and gangliosides (mundane molecules next to peptides, in his view). I was particularly interested, of course, to hear about the fate of scotophobin and its cousins, if any were on their way to revelation. He was philosophical about the insult of having his scotophobin paper to *Nature* held up for eighteen months, then published back-to-back with Stewart's rebuttal, which had so thoroughly nullified the scotophobin discovery. I used this opening to probe his view on dealing with rejection in science. All his career, he had been hounded by controversy, finding himself frequently outside the mainstream of thinking. Hadn't he felt frustrated to be thwarted so often? "No," was his response. "You have to remember that the ultimate aim of research is self-satisfaction."

A few days later, my aunt Virginia Rae had all the members of the family then in Houston over for an elegant Sunday morning brunch, laid out and leisurely enjoyed in the most gracious of Southern styles. As always before, the warm humid air of Houston had been like a balm to the raw and troubled edges of my being. This elegant touch added further to my regained composure. *How great it would be to live in Houston,* I thought, as the year inched toward its ending.

25

Arrival of the Endorphins

As 1975 ensued, Robin and I bore down on our two-track program in neurochemistry: me still determined to acquire convincing evidence of chemical gradients in the developing retinotectal visual system of chick embryos, and both of us slogging our way through an attempt to find functional plasticity in the metabolic profiles of brains from operantly conditioned pigeons.

Malaise in Mexico City

The ASN met in March 1975 in Mexico City. The meeting itself was completely overshadowed by its locale. There was a good session on gangliosides and a few isolated papers of high quality, but no overall theme predominated and no generalized excitement was evident. Most of us had come to visit Mexico.

I did take away one overall impression that related to my work with Robin. Never had I seen behavioral neurochemistry more in the doldrums than it was then. Not a single paper was presented on neurochemical changes correlated with functional changes in the brain. I had dinner one evening with Ed Bennett and asked him where he thought we stood with respect to biochemistry and behavior, and he said in essence that he didn't know and that he didn't think we were even very close to any new insights. He, like others in the field, didn't seem to know which way to go. Their favorite hunches had not borne fruit and, not having nurtured alternative ideas along the way, were at a loss of what to do next.[1]

Privately, I nurtured the thought that this malaise might be the opening that would benefit Robin and me. The time might be right for a breakthrough that the metabolic profile approach could provide. I thought that if we could carry out some key experiments that resulted in a few quality publications, we might yet be able to establish ourselves at the forefront of the field.

On a personal note, this was a good meeting for visitation. I had dinner with the Ungars one evening and a couple of good talks with Fred Samson. I also had fruitful discussions with Bob McCluer, Richard Quarles, and George DeVries, who asked me, as he always did, about Carol. They had shared a brief time together when both were at the Albert Einstein School of Medicine. Particularly gratifying was a brief conversation with Sam Barondes, who complimented me on a review I had recently published on glycolipids and glycoproteins in neural function.[2] He described it as "a monumental accomplishment," though he felt I had been a little generous with Bogoch, and he was probably right.

Storm beneath the Surface

The calm I had sensed in Mexico City was an illusion. Though not yet in the public domain, a fierce race was underway that spring to uncover the identity of the endogenous opiate that everyone agreed must be present, as the existence of a receptor for it implied. The three labs in hottest contention were led by Solomon Snyder, Hans Kosterlitz, and Avram Goldstein, respectively.[3] Those three labs updated their progress toward a chemical identity of one or more endorphins—a generic term meaning "morphine within" suggested by Eric Simon to include all the endogenous morphine-like peptides in the brain—at a meeting of the International Narcotics Research Club in May.

Gavril Pasternak from Snyder's lab had confirmed one amino acid, tyrosine, in an endogenous opiate. Hughes and Kosterlitz appeared to be closer to success. John Hughes had disclosed at a meeting chaired by Snyder at the Neurosciences Research Program the previous year that they had isolated a substance x, a peptide with an apparent molecular size of three to seven amino acids. Kosterlitz had named substance x enkephalin, meaning simply "in the head." They were now reporting that enkephalin consisted of at least four different amino acids in approximate ratios of three glycines, one phenylalanine, one methionine, and one tyrosine. Brian Cox from Goldstein's lab reported discovery of significantly larger peptides from the pituitary gland that were much more potent than the enkephalin preparations being used by Hughes and Kosterlitz.

Several circumstances had given the researchers in Scotland a slight advantage. First, they had negotiated with Barry Morgan at the Reckitt and Colman Drug Company a working agreement to process the massive amounts of pig brain needed for the isolation of the target compound. Secondly, they had turned to the use of high-performance liquid chromatography as an improved and much more sensitive method for isolating and purifying their candidate peptides. This enabled them to provide a peptide preparation in greater quantity and higher purity to Linda Fothergill, the biochemist who was doing the amino acid analysis and sequencing. And most critically, Howard Morris approached John Hughes after hearing the latter's lecture of enkephalin. Morris was a mass spectromotrist like Dominic Desiderio, who had helped solve the sequence of TRH with Guillemin and scotophobin with Ungar. Hughes was not enthusiastic about entrusting his valuable, limited supply of enkephalin to Morris, but stymied from further progress by the summer of 1975, he and Kosterlitz decided to let Morris see what he could do.

This was a wise decision. By mid-August, Morris was able to confirm the sequence of four amino acids: tyrosine, glycine, glycine, and phenylalanine, as Linda Fothergill had previously determined. Like her, though, Morris could not distinguish between methionine and leucine as the fifth amino acid. On a hunch, Morris told Hughes that enkephalin might actually be a mix of two similar pentapeptides—one ending in methionine, the other with leucine. When they chemically eliminated methionine from the mixture, only the variant with leucine remained. So there must be two variants: met-enkephalin and leu-enkephalin.

Synthetic met-enkephalin and leu-enkephalin were prepared as quickly as possible and, by mid-October, shown to be indistinguishable from the purified natural substances by Terry Smith in his bioassay using the vas deferens muscle. Knowing that Snyder and Goldstein were breathing down his neck, Kosterlitz requested priority treatment from the editor of *Nature*, who received his manuscript on November 13 and published it on December 18, 1975. There was no second-guessing the mass spectrometry evidence, no back-to-back critique this time, no eighteen months' delay. The announcement of "Identification of two related pentapeptides from the brain with potent opiate agonist activity" by John Hughes, Hans Kosterlitz, Linda Fothergill, Barry Morgan, Howard Morris, and Terry Smith was accepted at face value and honored as winner of the race.

In the meantime, Cox and Goldstein pressed forward with their research on peptides from the pituitary gland, which they claimed were many times more potent and long-lasting than enkephalin and which appeared to be over twice as large. In October, Howard Morris attended a lecture by Derrick Smyth at Imperial College in London on melanocyte stimulating hormone (MSH). Smyth was best known for his research on pro-insulin, a large inactive protein which contains the amino acid sequence of insulin, a breakdown product which has enormous physiological importance. Morris had begun to wonder if the enkephalins weren't breakdown products from a larger, inactive precursor, so Smyth's message resonated with him. The message became electric when Smyth's data revealed that the pentapeptide sequence of met-enkephalin was found inside the sequence of another part of the pro-hormone for MSH.

At this point, the scramble to discover more endorphins became crazy. MSH, it turned out, was part of lipotropin (LPH), a hormone that promotes the breakdown of lipids. Roger Guillemin had joined the scramble; and as soon as the enkephalin paper appeared in *Nature* with an appended paragraph suggesting the possibility of a relationship to

LPH, both Guillemin and Goldstein asked C. H. Li, the discoverer of LPH, to send them some of the purified hormone. Being a long-time friend of Guillemin's, but having been asked by Goldstein first, Li acceded to both requests. Li himself started looking for opiate activity in his samples of LPH, and Smyth, too, was doing the same after being alerted by Morris to the presence of the met-enkephalin sequence in LPH. Within months of the paper by Hughes, et al., in *Nature*, Goldstein and Li,[4] Guillemin,[5] and Smyth[6] were all claiming to have discovered new endorphins. The enkephalins were just the neuroactive tip of the endorphin iceberg.

In time, scores of new neuroactive peptides would be discovered while Ungar's discovery of scotophobin—the first apparent neuroactive peptide to be sequenced after oxytocin and vasopressin——receded into a footnote in the history of neuroscience. It would not be until a decade after the discovery of the enkephalins that a coincidental match between the first three amino acids of met- and leu-enkephalin and the last three amino acids of scotophobin would be noticed.[7]

Stood Up in New York Again

The drama of the endorphins was playing out in the background, as yet unbeknown to most of the three thousand[8] or so members who gathered for the annual meeting of the Society for Neuroscience in New York City in October of 1975. A few months earlier, Dianna had asked me if I might be interested in an open position in her new department at the University of Texas Medical School in Houston. "Of course," I responded, setting aside for the time being the knowledge that a move to Houston would entail a significant amount of persuasion for Carol to agree.

A couple of mornings into the meeting, I ran into Dianna, who asked if I would be willing to meet with her chairman, Joe Wood, for an informal interview. Again, the answer was yes, and after lunch, I got a message from her setting up a meeting for three that afternoon. I arrived at the appointed place at the appointed time, but Joe Wood did not—the Purchase experience once again! Only this time, I hadn't been forgotten, just preempted by an overly long session requiring Wood's presence, he explained the next day when, through Dianna's mediation, we did finally manage to meet. Our conversation went well. The department is in a state of flux, he said. All the new faculty are young, committed to neurobiology as a science, and looking for people willing to interact and collaborate. He asked if I were seriously interested in a new position, and I told him that, while I didn't have to leave Wayne State, I would seriously consider a position in his department because of the unique opportunity to get in on the early stages of the development of a group with great potential. I liked Joe Wood, and he seemed to like me. He said he would be in touch in about three weeks.

The meeting in New York also gave me a chance to see Penny. We had a couple of nice visits, and she gave me a ride to the airport the day the meeting ended. As usual, our time together had its poignant moments. En route to catch my flight, she decried how poorly her career had gone (it hadn't really) and how badly her personal life had messed her up professionally (though no one I knew had ever weathered personal storms better than she had). My last hour in New York, therefore, was spent trying to refute her overly harsh self-evaluation and cheer her up.

[1]   This pessimistic outlook on behavioral neurochemistry turned out to be premature for Bennet, Rosenzweig, and their colleagues, as within a few years, they would embark on what was probably their best, most sophisticated work. By using a mix of metabolic inhibitors and creative experimental designs, they dissected out the time course of metabolic events required for memory encoding in greater detail than anyone else. By then, though, the size of the audience that cared about and could appreciate their work had shrunk greatly from just a few years earlier.

[2]   Irwin, L. N. 1974. Glycolipids and glycoproteins in brain function. In *Reviews of Neuroscience*, edited by S. Ehrenpreis and I. J. Kopin. New York: Raven Press. Seymour Ehrenpreis had also taught briefly at the College of Pharmaceutical Sciences, where I had taught the pharmacology lab for his lecture course. He also knew me from the chapter I had coauthored with Ungar in the first series of review volumes he had edited. This had occupied much of my first few months in Detroit, when I was intent on establishing my credentials in the field.

[3]   Goldberg, J. 1988. *Anatomy of a Scientific Discovery: The Race to Discover the*

Secret of Human Pain and Pleasure. New York: Bantam Books.

4   Cox, B. M., A. Goldstein, and C. H. Hi. 1976. Opioid activity of a peptide, beta-lipotropin (61-91), derived from beta-lipotropin. *Proc Natl Acad Sci USA* 73: 1821–3.

5   Lazarus, L. H., N. Ling, and R. Guillemin. 1976. Beta-lipotropin as a prohormone for the morphinomimetic peptides endorphins and enkephalins. *Proc Natl Acad Sci USA* 73: 2156–9.

6   Bradbury, A. F., D. G. Smyth, and C. R. Snell. 1976. Lipotropin: precursor to two biologically active peptides. *Biochem Biophys Res Commun* 69: 950–6.

7   Wilson, D. 1986. Scotophobin resurrected as a neuropeptide. *Nature* 320: 313–4. The significance of this observation, if any, remains unclear.

8   I was overwhelmed and rather depressed by such a large meeting with so many scientists doing really impressive research. Little did I know that within decades, the annual meetings would explode to ten times that number of attendees.

26

Needle in a Haystack

By year's end in 1975, our research had reached a stage that could best be described as discouraging. Robin and I had worked very hard on the metabolic profile approach to learning in pigeons during the previous year but come up largely with negative results. The overwhelming dimensions of the task had finally sunk in, and our continuing inability to get grant support for technical assistance meant that we had very little chance of ever fairly testing the metabolic profile concept.

As I mulled over these discouraging thoughts, I decided to act on an inspiration that had come to me during the SfN meeting in New York a couple of months earlier. If what granting agencies want, I told myself, is absolutely cutting-edge, up-to-the-date methodology, with no particularly inspired theoretical framework, then I would resurrect my ideas (via Barondes and others) about the role of glycoconjugates in neural specificity during development and simply propose to use the fanciest, most esoteric methodology that I could find in the literature. So in the closing days of 1975 and the early part of 1976, I spent a lot of hours in the library.

The completed proposal was mailed to NIH in February. Given my terrible track record with NIH, I decided to submit essentially the same proposal to NSF—not that my success with NSF had been any greater. But having built a proposal around the strategy of extracting the zestiest methodology from the literature that I could find, the only strategy remaining was to spread the burden of rejection across as wide a range of agencies as possible.

Our Brief and Aborted Quest for a Cure to Cancer

Clearly, I was reaching a point of quiet desperation. One evidence of this was that Robin and I spent an inordinate amount of time during 1975 and 1976 applying to private foundations for the small crumbs of research money that their anonymous trustees—sometimes on the alleged recommendation of scientific advisory panels, sometimes not—dispense without explanation or accountability. Within about a six-month period, we tried the Dreyfus Foundation, the Anna Fuller Fund, and the Milheim Foundation for Cancer Research. My chairman, Walter Seegers, was kind enough to nominate me for a Sloan Foundation Fellowship. And Robin wrote an impassioned appeal to the Office of University Development at Wayne State for the Endowment of a "Developmental Neuroscience Research Program" (pleading, with characteristic fervor, that "the bureaucratization of science has coerced many talented and creative investigators into conducting applied and clinical research on cancer, heart disease, and a myriad of wounds and tumors. But science is a generalized abstract intellectual activity—it is a philosophy of discovery.") None of these efforts succeeded.

The Milheim proposal was a particularly interesting case. Carol had been working with Len Malkin in the Biochemistry Department on the biochemistry of tumors for some time when I came across a reference in the literature that CMP-sialyl synthetase, a key enzyme in the preparation of sialic acid for addition to glycolipids and glycoproteins, is localized in the nucleus of the cell. This is curious, since the ultimate destination for glycolipids and glycoproteins is primarily the cell periphery. Since the glycolipid composition of the cell surface was known to be different between cancerous and non-cancerous cells, it suggested the possibility that CMP-sialic acid conveys information from the nucleus to the cell surface, telling the cell whether to continue dividing or to cease dividing and become differentiated into its final form and function. This is the fundamental difference between normal and cancerous cells.

I drew up a grant proposal for a few thousand dollars to compare the activity of this enzyme (CMP-sialyl synthetase) in the tumorous and non-tumorous tissues that Carol was studying and submitted it to both the internal Faculty Research Award Program at Wayne State and to the Milheim Foundation for Cancer Research. Neither program considered it worthy of funding nor offered a critique, both responding with brief form letters of rejection. It was certainly a long shot at best, but I really believed that there might be a small chance that CMP-sialic acid serves as an intracellular messenger between the nucleus and the periphery of the cell. And were it true, it would have been a monumental discovery. But now I would never know. Overwhelmed by other duties and higher priorities, I was not in a position to pursue the idea without at least a small amount of funding. There is very little chance that the research, had it been pursued, would have turned up a cure for cancer. But the cost of testing whether the idea had any promise at all was not very high either.

Carol remained interested in the project and was willing to pursue it, but Len Malkin grew impatient with the diversion it had created for her and discouraged her continued participation in the project. The incident did have the side benefit, however, of reminding me once more that Carol and I had the makings of a good scientific team should we ever get the opportunity to work together over a longer period of time. Sooner than I realized and quite unexpectedly, that opportunity would arise.

Confronting Fears of a Logical Bind

The American Society for Neurochemistry met in Vancouver in the early spring of 1976, giving Robin and me an excellent reason for leaving behind the drabness of Detroit for a week. At the business meeting, we held a long and silly debate about whether to suspend the membership rules in order to admit Stanford Moore, a recent Nobel laureate, to membership without going through the usual channels. This motion was pushed by one particularly loquacious member who appealed to our need for stature in the Society. As one who had suffered what I considered an unnecessarily grueling test for admission, I clearly was in no mood to suspend the rules for anyone short of Louis Pasteur, and the rest of the membership agreed.

Robin and I presented a poster on our pigeon brain–metabolic profile research. It was about the only work presented at the meeting that had anything to do with learning in an intact animal. This fact was commented upon and lamented by Ed Bennett, who—though he personally had no direct, vested interest in the outcome of our experiments—was beginning to see us as among a diminishingly small circle of neuroscientists like himself who were still willing to look for chemical correlates of learning, however remote the possibility of success was beginning to look to others.

Notwithstanding the relative lack of data on the subject at this meeting, there was a symposium on the neurochemistry of learning and memory. Bernard Agranoff, in a valedictory lecture at the end of his term as president of the Society, offered a commentary that articulated a growing body of thought at the time and even struck a disturbingly responsive chord deep in my own mind. He began by noting that searching for neurochemical correlates of learning is like searching for a needle in a haystack. The real problem is not knowing either the dimensions of the needle or the haystack. This I fully agreed with. But then he took the next step into an intractable logical bind that I had often sensed but never allowed myself to accept: If one *does* find an apparent chemical correlate of learning, it is unlikely to have anything to do with learning because finding a true, *isolated* correlate of learning is in reality so improbable.

I then offered from the audience a commentary to his commentary. Citing our finding that about 10 percent of all morphological-metabolic compartments in the pigeon brain showed some response to behavioral treatment, I asked if that constituted a believable magnitude for the needle. His response was a vague but orderly retreat from the logical bind into which he had cast us all. But a vision of the future of behavioral neurochemistry had been laid down. One of the world's most influential neurochemists had declared that the discovery of true neurochemical correlates of learning was, at that time, a virtual impossibility.

Nightcap

At Robin's urging, we retreated for consolation to the hotel bar after dinner that evening. As it happened, there was a very attractive Air Canada flight attendant unaccompanied among the bar's few other customers. It didn't take long for Robin to switch from consoling me to courting her. We were soon a threesome, drinking, talking, and eventually dancing. At first it was only her and Robin, as I protested that I was a lousy dancer. In all such social interactions involving both Robin and me, he naturally dominated the situation, and this was no exception. Thus he totally consumed her attention at first. As the evening wore on, however, she did gradually talk to me more, and at the urging of them both, she and I danced a time or two. We talked about her job and then about our job as college professors and neuroscience researchers. She professed to be interested and impressed and apparently decided as the night progressed that we were nice guys, even if we were from Detroit. More importantly, she proclaimed that I wasn't a bad dancer after all.

Finally, we were chased from the bar, and Robin invited her to our room. By this time I was very tired and, having been declared an adequate dancer, willing to call the evening a success. Robin was not so easily satisfied, and when she accepted his invitation to join us for a nightcap, her intentions became unclear to me as well. So she and (mostly) Robin continued to talk in our room as the night got later and I got sleepier. Finally, she offered to leave, and Robin kindly offered to walk her to her room. She said it wasn't necessary, but he insisted it was no imposition.

The next morning, Robin told me that they had just continued their conversation and that nothing in particular had happened between them. He told me this with a frustrated edge to his voice. Years later, when on occasion we would start to talk about that night in Vancouver, he would smile briefly and change the subject.

27

Need for Another Moving Van

Back in Detroit, the annual Physiology Department meeting approached. Walter Seegers, the chairman, had an aversion to departmental meetings, so restricted them to a single three-day marathon in the spring, during which we all took turns congratulating ourselves on our various successes, electing ourselves to various committees, and discussing those few policy issues about which the chairman was willing to entertain our opinion. Particularly because of the requirement for self-posturing, Robin and I did not enjoy this event. While others could point to their recent grants and complain about their inadequate number of technicians, we had to confess to another year of keeping our research going at a subsistence level. Fortunately, within the past year, we had finally published or had accepted for publication three papers, and it was generally acknowledged that no one worked harder at research than Robin and I; so our colleagues were sympathetic. But by comparison, our accomplishments seemed meager.

There was one bright spot to the meeting that really touched us. The faculty voted to make indirect cost rebates to the department available to those members of the department (Robin and I) who did not have any outside research grants. Every outside grant includes a budget item for indirect costs or overhead; and Wayne State, like many research institutions, rebated a fraction of these funds to the departments that generated the grants, obviously as an incentive for them to produce *more* grant supported research.

Much is made in the literature on the sociology of science about the competitiveness of science and the egocentricity of scientists. And much of it is true. But throughout my career, I have seen many acts of generosity, selflessness, and honor among scientists as well. I was fortunate to be at Wayne State at a time when it was an aspiring research university but, at least at the faculty level, still capable of a collective attitude and a generous spirit. Robin and I, who would be the main beneficiaries of this policy, were humbled and touched.

Getting a Lifeline from NRP

Now the summer approached and the time for a decision on my tenure grew near. Back in the winter, I had been reasonably confident that I would be granted tenure: Robin and I had publications coming out, my teaching was excellent, and the recommendations from my department and from Walter Seegers had been strong. As the new year wore on, however, my confidence began to slip. Part of it was fallout from the growing doubts I had about the direction of my research, and part of it derived from vague comments made by Robin, who was always well plugged into the rumor mill. In retrospect, he seemed to be warning me to brace myself for a negative decision.

On the morning of June 9, when I arrived back at my office following my eight o'clock physiology lecture, a note for me to see Dr. Seegers was waiting. When I walked into his office, the issue of tenure was not on my mind, so when he asked if I knew what he wanted to see me about, I said I had no idea. Just as I said that, though, I saw he was holding a piece of paper inscribed with

> Irwin –
> Murano –
> Walsh –
> Sedensky +

I knew immediately that Ralph Walsh, Gene Murano, and I had been denied tenure; or in Seegers's direct and inelegant way of putting it, "We bombed out on this tenure business."

My first reaction was to be impressed with and touched by the fact that Seegers seemed genuinely dismayed by the news that he had to convey to me. At some level not so deep in my psyche, I had prepared myself for this outcome. That, too, deflected the blow. Nonetheless, it was another rejection, and one of mammoth proportions this time.

For the second time in less than four years, I was going to have to fight for my professional survival. Having weighed the possibility in recent weeks that the tenure decision might go against me, I had already made the decision that I would not appeal it should it be negative. I had watched others go through the humiliating grind of the appeal process and vowed to avoid that indignity. I was not so enamored of either Wayne State or Detroit that I would fight to stay where I was apparently not appreciated. Instead, I immediately put into action my fallback strategy: I wrote to Frank Schmitt at the Neurosciences Research Program in Boston, asking if there were any openings for staff scientists in that organization.

As soon as he received my letter, Schmitt called me with apparent enthusiasm, relating NRP's latest grandiose plans for a new Intensive Study Program (ISP) being scheduled for the coming summer in Boulder and asking if I expected to be in the Boston area anytime soon. I told him that I had no plans to travel to Boston but could easily make them, whereupon he invited me to come for an interview, which I did on June 24. Of the people I had known at NRP during my work there in 1968 and 1969, Frank Schmitt (Director), Kay Cusick (Executive Assistant), and George Adelman (Librarian) were still major figures, though Schmitt was now a Foundation Scientist and Cusick was Associate Director of Business and Finance. Replacing Schmitt as Director was Fred Worden, a neuropsychologist formerly of UCLA, whom I had not previously known. Barry Smith, one of Schmitt's last graduate students, had become Program Director.

The job for which I was interviewing was Staff Scientist, an ill-defined position that, for years, had defied anyone's ability to characterize with accuracy. Basically, the staff scientists at NRP carried out background literature reviews, in-house writing assignments, and low-level organizational tasks as delegated on occasion by Schmitt, Worden, or the Program Director. In the elevated prose of NRP grant proposals and progress reports, they carried out "conceptual research," but their ability to affect policy or the scientific direction of programs was strictly limited by the whims of Schmitt and Worden. Staff scientists were, in fact, primarily a technical staff responsible, first and foremost, to Frank Schmitt on a deep background basis.

At the time, I did not fully understand the weaknesses of the position, though I certainly appreciated that NRP was strictly a Frank Schmitt show. Instead, in my haste to see NRP as a quick and honorable means of disengagement from Wayne State, I focused on the possibilities inherent in the rhetoric of conceptual research. Though Worden did not know me, Schmitt and Adelman remembered me well for the salvation of the "Brain Cell Microenvironment" work session that I had achieved during my summer stay in 1968 while still a graduate student. And Kay Cusick, who had obviously acquired more influence, seemed to remember me fondly. Barry Smith also apparently had favorable recollections of our interactions at the 1969 Boulder ISP. So after a huddle of not more than ten minutes, Worden ushered me into his office and offered me the position of Staff Scientist at $19,000 a year, a thousand more than my then current salary. By day's end, I was flying back to Detroit with a job offer from MIT (the parent institution of NRP). Dr. Seegers was delighted that I had come up with a prestigious alternative so quickly. I waved it in the face of the higher administration briefly, notifying the dean that Wayne State could retain me if they would meet MIT's salary. When I got no takers immediately (as I knew I wouldn't), I hastily dispatched a letter of resignation the day before I received official notification that my application for tenure had been denied.

Trip to the Sand Dune

Back in the spring, Robin and I had conceived of the idea of a research retreat, whereby he and I and all the students working in our lab, and their spouses, would retreat for a few days in June to a cabin on the eastern shore of Lake Michigan to map out our work for the coming summer and fall. To parody the hype over the nation's bicentennial that year that the Ford Administration was promoting with great show and little substance, Robin and I dubbed our excursion the First Annual Bicentennial Neuroscience Research Retreat and Campout. Marc Abel, who by then was deeply involved as a graduate student on peptide projects with Robin, and his wife Lori, had access to a cabin a little north of Petoskey, Michigan. Our notion was for Robin, me, and the students to meet during the day to talk about our various research activities while our families played, then in the evenings, we would all relax and socialize.

By the time the appointed week had arrived, however, several students threatened not to go, whereupon Robin backed out. However, Carol and I were very much in the mood to get out of Detroit. David Terrian and Jon Klauenberg joined Marc in forming a small cadre of students who were anxious to get out of town as well, so they and their wives and Carol and I with Anthony headed north for a week of togetherness in the country.

Four of the seven days we were there were sunny and beautiful, so we spent a lot of time on the beach about a hundred yards from the cabin—a picturesque walk down the hill and through the woods, past a small sandy bluff that Anthony dubbed the sand dune. Each family took turns fixing the evening meal, and a friendly competition developed among us to see who could prepare the more memorable feast. After a day on the beach and an afternoon of volleyball, Suzanne Klauenberg would fix us all a strange drink consisting of one part burgundy to three parts ginger ale—my first exposure to what would later develop into the huge commercial success of the wine cooler. After dinner, we would retire to the living room with its huge piano, where Marc would play a long string of mostly his own musical compositions, displaying the talent he would later turn to writing music for the off-off-Broadway stage as a sideline to his research in suburban New York. Forever after, whenever any of us would refer to that brief but memorable vacation, we would speak of the trip to the sand dune.

What made it memorable was the incredibly positive chemistry that developed between the four families—we got to know one another as real friends outside the context of the classroom and lab—combined with the melancholy realization that my impending departure from Wayne State would leave these friendships undeveloped. Fortunately, it turned out that over the years, Carol and I were able to keep in touch with Marc and Lori Abel and Dave and Carrie Terrian, as Marc and Dave both completed work on their doctorates and moved on to accomplished careers in teaching and research in neurochemistry.

While at the sand dune, Dave and I mapped out what was to be my last big experiment in Detroit: extraction of brain gangliosides from rats subjected to various combinations of hormonal manipulation (by adrenalectomy and adrenal steroid replacement) and behavioral manipulation (by the swim-escape procedure first used in my doctoral work, but nicely developed further by Jon Klauenberg at Wayne State and later made famous in slightly altered form as the Morris water maze). Dave did a great job of surgery on the rats, and the adrenalectomy/swim-escape experiment turned out to be perhaps the highest quality of research that a student and I did in Detroit. Unfortunately, the results were not compelling; gangliosides, like nearly every other molecular endpoint on which Robin and I had focused, showed a disappointing lack of responsiveness to the various internal and external stimulations that we were providing. It became clear that our level of resolution was not great enough to answer the questions we had posed. The needle really was too small for the size of the haystack, and we simply weren't going to find it with a tool as crude as a pitchfork.

Nights in Ontario

Now, in the early autumn of 1976, my career again hung by a thread when a phone call from Washington intervened. A program director from the National Science Foundation called me on November 1 to inform me that he wanted to fully fund my grant proposal on topochemistry of differentiating nerve cells. My hours in the library had paid off. Finally, after years of effort and a multitude of failed grant proposals, I had finally crafted a winner.

The irony of the situation immediately set in. Having failed to gain tenure due to lack of external funding, I now had a grant but no job with which to execute it. And transferring the grant to NRP in Boston was not a solution, since NRP had no laboratory facilities (none were needed to do conceptual research) and no inclination to let its staff scientists develop much of an individual effort or identity. Furthermore, Robin was a coinvestigator on the grant, and taking it with me necessarily would deprive him of its benefits; and in his untenured situation, that hardly seemed fair. So the good news from Washington brought with it a sizeable dilemma, as Robin and I, and all our graduate students and their spouses, prepared for the Sixth Annual Meeting of the Society for Neuroscience Meeting in Toronto—a meeting destined to be memorable for more reasons than one.

For starters, it was clearly the year of the peptide. The entire scientific establishment was in a buzz of excitement over the discovery that endogenous peptides bind specifically to opiate receptors in the brain. It was also a time at which the hippocampal preparation, a thin slice of cerebral tissue that contains a sequential relay of several neurons whose activity can be potentiated or altered by recent experience, was coming to the fore as a possible model for learning in the central nervous system. Gary Lynch, one of the rapidly rising superstars at the University of California at Irvine, gave a talk on long-lasting potentiation in the hippocampus, while Timothy Teyler, who had introduced the hippocampal slice preparation to this country from Scandinavia, gave a characteristically more modest poster on the same subject. Oswald Steward, another pioneer in the hippocampal field, presented or coauthored three papers on the subject. And Samuel Barondes got into the act by talking about membrane fluidity and synaptic potentiation.

Retinotectal neurospecificity was still the rage. A host of young scientists reported on splitting, compressing, expanding, and ablating both retinas and tecta in ever more convoluted ways with the usual confusing results, though the spectacle of Martha Constantine-Paton and her three-eyed frog was missing. Bernard Agranoff, as if to shadow my career, presented an in vitro model for retinotectal connectivity in the goldfish. As usual, there was little chemistry in evidence, but Gerald Edelman, the Nobel Laureate of past and future exploits at NRP, did present with Urs Rutishauser and other students a paper on cell adhesion in the chick embryo—a phenomenon that in time would turn out to depend on a specific family of glycoproteins.

This in many ways was the most relaxed and enjoyable neuroscience meeting for me to date. Robin and I did not present a paper or poster, so we didn't have to worry about that; but knowledge that I had received a grant from NSF for my particular approach to retinotectal neurospecificity allowed me the luxury of unconcern about the plethora of experimental manipulations of that system being reported. My only real discomfort at the meeting came in my conversation with Georges Ungar, in which I had to tell him about my impending move to NRP. There was no love lost between Ungar and NRP to begin with, as he felt they had snubbed him when they could have turned the neuroscientific world in his favor. In addition, he had no particular respect for conceptual research and thought it a waste of my talent to be going in that direction. I understood his cynicism and shared it to a degree, but explained that I had no choice, given the situation for me at Wayne State and that it would probably be temporary only (not knowing in truth whether it would or not). I put the best face on it I could, but knew that he was disappointed in me, and I hated that.

On the other hand, Fred Samson and Dianna Redburn were both encouraging.

Fred, of course, was, to a substantial degree, a creature of NRP. We had a good talk in which he advised me how to deal with NRP and what to do about my grant, giving me ideas of different labs in Boston where I might carry out the research. Dianna, radiant in her new role as mother and weighted down with a pound or two of photos of her baby boy, was encouraging. While not saying it in so many words, her attitude was clearly, "If anyone can do conceptual research, you can." Especially given the fact of my new grant, she had a feeling "that everything was going to work out well" for me. I soaked up the encouragement, anxious for every tidbit I could absorb.

One evening, I decided to absorb some culture. Usually, I skip evening sessions at scientific meetings since, given the hour, they tend to be either too boring or too volatile. But one of the special interest dinners at the Toronto meeting was scheduled to be held at the site of the McMichael Canadian Art Collection in the village of Kleinberg, some distance outside of Toronto in rural Ontario. We were transported by bus to the museum, where we were to have a guided tour of the art collection, followed by a pioneer Canadian dinner, then by a scientific program on developmental neurobiology. We made it to the museum more or less on time, despite the onset of a significant snowfall. The tour was fascinating; the art (much of it native art, featuring the works of Canada's famous Group of Seven) was beautiful and the setting perfect. The wine and spirits made available during our cocktail-hour tour sharpened our aesthetic appreciation as the evening wore on.

Unfortunately, the deteriorating weather apparently made delivery of food to our remote location difficult, for the cocktail hour drifted into the dinner hour with a continuing flow of wine and spirits, but not appetizers, entrées, or caffeine to help us keep our wits about us. None of this would have mattered, probably, had not the chairman of the scientific program been Jack Diamond, a professor of Neuroscience at McMaster University in Toronto. Garrulous, articulate, and witty, Diamond believed that having a good time is almost as important as having good science; and on this particular evening, the elongated cocktail hour strengthened his commitment to showing his guests a good time. As the hour grew later without food, Diamond regaled the company with a barrage of stories and witticisms that grew in length and humor as the evening and the wine wore on. At some point, that threshold was crossed where most of us lost both our zeal and capacity for getting serious about developing neurons. Being the gracious host that he was, Jack put up to a vote whether we should proceed with the scientific presentations or call it an evening. The minority who still wished to talk science were booed down by the majority of us who were in no condition to take anything seriously anymore, so we loaded the bus for the return to Toronto, imbued with rural Canadian culture but no wiser in the ways of developing neurons (having doubtlessly killed more than a few of our own).

That was not to be the only memorable evening for me in Toronto. One night, after a long day of science and socializing, I was almost asleep in our room at the Sheraton Crown Center, when I heard the door open, followed by muffled voices belonging to Robin and an unidentified woman. Seeing me obviously sleeping, they kindly slipped into the bathroom to continue their conversation in a manner that would not disturb me. I was almost asleep again when I heard the sound of running water. Robin had decided to take a bath, his companion having departed, I assumed. Within a few minutes, however, I began to hear the sound of water sloshing with vigor in the tub, accompanied by muffled exclamations that were not exactly groans but not exactly conversation either. Now thoroughly awake, I concluded that Robin's friend had not departed. In time, the noises subsided, the bath apparently concluded; whereupon, Robin and his friend emerged into our darkened bedroom. Robin walked up to my bed to see if I were indeed asleep, which I certainly was not, though in due consideration of his privacy, I feigned to be as near unconscious as I could under the circumstances. Not wanting to be rude, he called his friend to my bedside and introduced her. I looked up hazily at the silhouette of a shapely young woman wrapped in a bath towel and shook hands with her, as she greeted me most cordially and apologized for the disturbance. No problem, I assured them both and turned over to let her get dressed and say good night to Robin in simulated privacy.

That was not their plan. Robin, having apparently been assured that their presence was no problem for me, ordered an erotic movie from the television, then quickly climbed into the adjacent bed with his friend. There followed an hour or two, at least, of the most ardent and artful (and only) lovemaking that I had ever been privileged to witness in person. While I feigned sleep as well as I could during the entire episode, anyone who could have slept through that show would have had to belong to another species. When at last the activity subsided, Robin and his friend fell into a relaxed and blissful sleep, leaving me anything but relaxed and unable to sleep for another two hours. At last, I drifted into a semiconscious daze, till the sound of slightly bustling activity in the room woke me. By the faint light of early dawn, I saw our attractive guest emerge from the bathroom fully dressed, give Robin a quick kiss, and thank him in brief but earnest terms, then discreetly slip out the door.

Later in the day, as I sleepily cruised through an aisle of posters in the exhibition hall, I came upon her; but she did not recognize me, and I managed to suppress an impulse to introduce myself. The only thing I ever knew about her was that she was a graduate student in neuroanatomy somewhere in North Carolina. She is one of the few people in this story whose fate I never learned.

The Long Goodbye

My teaching duties during my final semester at Wayne State were unusually heavy, but particularly gratifying, perhaps because I poured by body and soul into teaching as if to say to Wayne State, "Look at what you're willing to throw out, for the sake of a grant that you didn't have the patience to wait for." At the end of my section of the anatomy and physiology course, I received a round of applause that touched me deeply.

If my teaching was driven by a desire to embarrass the university's poor decision in letting me go, my research was motivated by a desperate attempt to accomplish something substantial in the few weeks that remained to me in Detroit. For it was distinctly possible that this could be the last laboratory research I would ever do. Dave Terrian had done a great job with the complicated experiment on swim-escape learning in adrenalectomized rats that we had designed at the sand dune back in the summer. Marc Abel and Jon Klauenberg were working more directly with Robin and making good headway toward their degrees. There was one part of another experiment that Robin, Dave, and I were working on that demanded the processing of subcellular fractions from a large number of regional brain samples, a chore that fell mainly to me. Shortly before Christmas, I pulled the rotor out of the centrifuge that had separated the last subcellular fraction from the last brain sample, as someone waited behind me to use the centrifuge for another experiment. "I'll be out of your way in just a minute," I said, then under my breath to the wall, ". . . possibly forever."

The reason for my uncertainty lay in the fact that, technically, federal grants go to the institution, not the investigator. They presume that the work will be carried out and administered by the institution from which the proposal is submitted. The National Science Foundation and the National Institutes of Health have always been lenient in transferring grants when scientists move from one institution to another, but it is by no means automatic. And my case was complicated by the fact that my coinvestigator, Robin, was not moving with me. He could, in principle, keep the grant at Wayne State; but the actual work that we proposed to do was much more closely dependent on my expertise, and I had in fact written the proposal. So Robin, with no apparent hesitation, agreed unconditionally to relinquish any claim to the grant and to support my request to NSF that the grant be delayed until I got settled in Boston, then transferred to an appropriate lab there where the work could be carried out. It was a singular act of generosity that, though not atypical for Robin in his relationship with me, was all the more meaningful because, as an untenured assistant professor, he was as vulnerable as I, yet he was giving up one of the main arguments that he could use in seeking tenure the following year.

Two possibilities emerged as sites to which I conceivably could have the grant transferred in Boston, one of which was the Eunice Kennedy Shriver Center for Mental Retardation in suburban Waltham, where Bob McCluer was head of the Biochemistry Division. Bob and I were friends of long standing, by virtue of our mutual interest in gangliosides.

At Thanksgiving, Carol and I flew to Boston to look for a place to live and to visit the Shriver Center. Our interview with Bob McCluer went splendidly. He seemed genuinely enthusiastic about recruiting us—both Carol and me—to the Shriver Center. Carol, who had met him several years earlier at the Hershey meeting of the ASN, liked him and felt good about the facility. So pending official approval from both NSF and the administration at the Shriver Center, it would be that lab in Waltham where I would make a stab, this time in partnership with Carol, at salvaging my research career.

The time for goodbyes was finally upon us. The festivities began with a full-scale Christmas dinner on December 12 at the home of Dan Michael, an undergraduate who had been drawn into our circle through his uncle's attendance at our evening neuroscience seminars. He, his mother and sister, and girlfriend Melinda prepared a sumptuous feast for Carol and me, Marc and Lori Abel, Dave and Carrie Terrian, Jon Klauenberg, and Katie Frank. This same group assembled once again at Jon and Suzanne Klauenberg's apartment a week later for a wine-and-cheese party at which they presented me with a lithograph by Peter Max. Knowing my love of art (to Carol's dismay, I spent too many Sunday afternoons running around Detroit to art auctions), they had pooled their resources to purchase a work showing the spare profile of a face with two small people—one sitting on the forehead in a posture of meditation, the other standing back watching. I told them the first thing I would do in Boston was hang it on the wall of my new office. To this day, it is the first thing I have always hung in a new office wherever I have moved, because that group of students, like the ones I left behind over three years earlier in New York, had come to occupy a special place in my memory.

The Physiology Department at Wayne State had a very nice luncheon for Carol and me. Dr. Seegers, my chairman, as always was gracious and made me feel that he was genuinely sorry to see me go. Also, Marilyn and Jerry hosted a gala neighborhood party for us—a fitting end to our involvement in the multiracial inner-city neighborhood of Detroit that had been our home for three and a half years.

Finally, in our last night in Detroit, just Marilyn and Jerry and Carol and Anthony and I had our final dinner together. This was surely the hardest bond to break, so interwoven had our lives and careers become since September of 1965 when Jerry had taken me under his wing as a first-year graduate student at the University of Kansas and since the late summer day in 1966 when Carol and Marilyn had first met sight unseen as roommates thrown by chance together.

The next day, we headed out of Detroit for our second attempt to find a permanent home and place to work on the East Coast.

LATE SEVENTIES (BOSTON)

28

Grasping for Silver Linings

From a southerner's perspective, there are a lot of things wrong with Boston. The natives are curt and uncordial. The neighborhoods are balkanized, with some as racist as any in the South. The traffic is dense, and the drivers are abusive. The roads are crooked, the intersections unmarked, and the streets narrow and too often unnamed. The beaches are more rocky than sandy, and the water that washes them is frigid and unclean. Most of all, the winters are cold, wet, and interminable. But for all its faults, Boston is a place that matters. Like New York City, the Beltway around Washington, and the coast of California, important things happen there. Especially in the realm of academic science, there is probably a greater quality and quantity of research per unit area carried out within an hour's drive of Massachusetts Bay in any direction, than anywhere else in the world. To fully taste the academic flavor of America, sooner or later, a scientist has to spend at least a little time in Boston.

I had to keep reminding myself of this during my first weeks in Boston because they were pretty miserable. We arrived on January 2, 1977, during one of the coldest, snowiest winters in several seasons. We had ended up leasing a condominium in the little town of Stoughton, nearly twenty miles south of the Brandegee Estate where NRP was located near the Jamaica Plain section of Boston. There was no simple way to get from that part of Boston to Stoughton, and the first couple of weeks of my new position were spent in experimenting on how to find NRP every morning and how to get home every night. The evening commute was particularly frustrating. It was always in the dark, since nightfall comes about four in the afternoon in January in Boston. The excessive snowfall that winter had narrowed the streets even further and obscured many of the infrequent street signs, so I creeped alternately through Hyde Park or Roxbury or Roslindale every evening, hoping against the odds that I would not get lost, but usually doing so anyway. One particularly frustrating evening, I ended up at the Rhode Island border, having taken a wrong fork thirty miles earlier with no idea of how to retrace or correct my errant path. On one frigid afternoon, I got stuck in a snowbank, and after half an hour of digging snow out with my foot, I went into a senseless rage of frustration, stomping at the tire and cursing my banishment to the frozen tundra of New England.

Conceptual Research and the Chain of Command

My disposition was not helped by the fact that I had to spend the first few weeks trying to figure out what it meant to be a conceptual neuroscientist. The early signs were moderately encouraging. In the small, cold corner office assigned to me, my first task was to pore through folders on an upcoming work session on the influence of light-dark cycles on neural biorhythms. Fred Worden had dropped the material on my desk one morning and asked for a boil down (the NRP term for a short synopsis of the various points of view contained in the publications and correspondence on the subject) by that afternoon. I managed to produce a page-and-a-half report for him by three thirty, with the distinct feeling that this task was a probe of my ability rather than a substantive assignment. He professed to be pleased with my performance on this first test.

Thus emboldened by my first positive experience, I decided to write a letter to Victor Shashoua. He had written to NRP suggesting that a work session be held on the biochemical basis of learning and memory.

I had found his letter in going through the files and had decided to answer it on my own initiative. I told him that I supported his call for a work session on the biochemistry of learning and memory and would work for it internally. I warned him not to overestimate my importance at NRP and discouraged him from taking this as anything more than a personal commitment. As usual, I dictated the letter to Marie Foley, the secretary for all the staff scientists and, after editing the rough draft, gave it to her for final typing. The following morning, I was shocked and amazed to find a copy covered with editorial changes in red ink that had the effect of turning it into a totally bland and meaningless letter, with a note from Fred Worden asking to speak to me about it. I first went to Marie and asked her how he had gotten a copy of my private correspondence, and she apologetically told me that she had standing instructions to clear all correspondence from staff scientists that implied a commitment by NRP with Schmitt or Worden. I then went to Worden, and he was gracious in explaining that NRP had been burned in the past by apparent commitments made by staff members that the organization had not been able to uphold. I took his point and understood the rationale, but resented not having been told the policy beforehand. As an academic spoiled by the luxury of autonomy throughout my career to that point, I was crushed by the indignity of having my personal correspondence censured. As I grew older, I came to realize that chains of command have a place, even in academia. At the time though, it simply reminded me of the low status to which my job (I began to think of it as a job, rather than a position or appointment) had sunk.

I grasped for silver linings. George Adelman, the NRP librarian and devoted fan of mine since my salvage of the "Brain Cell Microenvironment" work session in 1968, became my close confidant and good friend again the day I walked into NRP. Unexpectedly, another bright spot came in the form of the Program Director, Barry Smith. My first acquaintance with Barry had come at the 1969 Intensive Study Program, where his contribution to our staff work had struck me as weak and unimpressive. When I had interviewed for the staff scientist position at NRP the previous August, however, Barry had impressed me as a completely different person. Gone was the passive and deferential appendage to Schmitt, replaced by a confident, new individual, with his own thoughts about the mission of NRP. He stated his belief that NRP was going to have to change to remain viable—that what it had done in the past was useful and good, but no longer unique. Toward that end, he was quietly assembling his own network of younger advisers and seeking to find new ways for NRP to impact the neurosciences. Since these thoughts all matched my own much better than anything that Fred Worden or Frank Schmitt had to say, I was greatly encouraged.

There were two other staff scientists at NRP when I arrived in January of 1977. Parvati Dev had been trained as a theoretical neuroscientist and had interacted very well with Frank Schmitt. He had come to rely particularly on her and Barry as planning for the upcoming ISP had progressed. They had coauthored an article in science that was a fairly creative rehash of points made years earlier by Ross Adey, Ted Bullock, Gordon Shepherd, and others about the importance of local interactions in neuronal circuits—the central theme of the ISP planned for later that year in Boulder.

The other staff scientist was Bruce Brandt. He had worked with one of NRP's favored Associates, Maxwell Cowan, and had a level of irreverence that I found helpful and refreshing. Largely because of that, he was not favored by the major players at NRP. He and I got along well, though, and I admired his independence.

It was Key Dismukes, however, with whom I would eventually form the closest bond. He did not arrive until March of that year, recently from an appointment at the Institute for Ethics, Society, and the Life Sciences at Hastings-on-Hudson in New York. Prior to that, he had held postdoctoral positions with John Daly, a neurochemist at NIH, and with Sol Snyder at Johns Hopkins. He had, in fact, overlapped somewhat with Candace Pert as she was getting on to the research track with opiates that would make her famous. As a graduate student at Penn State, he had worked on the neurochemistry of learning and memory without much success—an experience that had left him with a thorough skepticism about the field. For this reason, he provided a healthy counterpoint to my generally more positive attitude. His social graces were inferior even to mine; but he was a gifted writer and, like me, held in disdain the attitude prevalent at the top of NRP that truth and wisdom were the sole domain of neuroscientists who had attained a degree of eminence in the eyes of Frank Schmitt or Fred Worden.

In mid-February of that year, the harsh winter turned suddenly benign. After Key arrived and the temperature began to climb, so did my spirits and hopes that something might be made of conceptual research after all.

Students and Mentors

In March, I requested and was granted travel funds to attend the 1977 meeting of the ASN in Denver. I managed to arrive on a Friday night on one of the first flights to get in after a major snowstorm, which left the ski slopes west of Denver covered with fresh powder snow and relatively devoid of skiers. I took the opportunity to ski in the Rockies for the first time, doing fairly well until a midafternoon banana daiquiri left me uncoordinated enough to turn my final run into a succession of multiple tumbles all the way down the mountain.

Marc Abel and Dave Terrian, our students from Wayne State, were attending to present their first papers at a national meeting. The first couple of nights we spent in rehearsal. Marc was nervous but clearly up to the task. It wasn't clear that Dave was equally confident. Despite numerous rehearsals, Dave had a rough time with his talk on our multiple variable experiment that had combined adrenalectomy, corticosterone replacement, and swim-escape behavior in an effort to determine if either endocrine or behavioral manipulations would influence ganglioside metabolism (for the most part, they didn't). Finally, he asked me if I would give the talk instead of him, but I said no, fearing that such a failure might be more than he could subsequently overcome.

Marc was more fortunate in that his presentation was scheduled for Tuesday, the second day of the meeting. His talk on rat brain peptides went well. Dave had to wait till Thursday for his talk, adding two days of agonizing anticipation that failed to clarify his understanding of the work well enough to give him any confidence. So he gave a stumbling, robotic, presentation; but he did get through it. Twelve years later, at an international meeting in Portugal, well after Dave had received his doctorate and established himself as one of the nation's premier young researchers on the neurochemistry of the hippocampus, I sat in an audience of several hundred and listened in awe to a confident, self-assured, tour-de-force presentation on the innovative research he had developed from such a halting beginning back in Detroit.

On Wednesday evening, I met the Ungars (Georges, Alberte, and their daughter, Catherine) for a drink at the cocktail hour. The moment I saw Georges, I realized that the rumors about his health were true—he had contracted cancer and was gravely ill. He was gaunt and thin with all remnants of hair gone. But he brightened when I appeared and really seemed happy to see me. He was particularly pleased when I told him about my NSF grant and that I would be restarting my research at the Eunice Kennedy Shriver Center that summer. Before long, he excused himself in order to rest before a panel discussion he was later scheduled to take part in. After he left, Alberte filled me in on details of his condition, adding, "As you know, he is very fond of you."

The panel discussion provided a sad conclusion to the day. Ungar defended the behavioral transfer experiments as a form of bioassay, but detractors in the audience were not in a compromising mood. It was clear that several younger, aggressive attendees knew nothing of the history of the field, but that didn't inhibit them from expressing their hypercritical and uninformed opinions. Two nights earlier, a similar scene had played out in a panel discussion led by Adrian Dunn, whose work was shedding important new light on the role that peptide hormones play in modulating memory storage and retrieval processes. But the relationship of peptides in any form to behavior in any way had become so tainted that the audience was clearly unreceptive. This was all so strange in light of the opiate peptide mania that was sweeping the world of neuroscience at the time. But the fact was that the opiate craze had brought a lot of new, young people into neuroscience who had little knowledge of the backstories, and those who *had* read the literature were probably impressed with the apparent disarray into which the field of peptide research had fallen.

Marc and I skipped the last morning of the meeting to drive up into the mountains and bask for a time in the beauty of the Rockies before flying back eastward to uncertain futures. He was headed to Houston, where he would later earn a doctorate at the University of Texas Health Science Center, working in part with Dianna Redburn. And I was returning to Boston, to the privilege and frustration of trying to define and accomplish something meaningful at NRP. Much on my mind, though, was the melancholy sense that the flurry of work and excitement on neuroactive peptides was leaving Georges Ungar unvindicated and underappreciated.

29

"Edelman Is No Neurophysiologist. He Needs Mountcastle!"

Back in Boston, I wrote a lengthy memo on the high points of the Denver meeting for distribution to the science staff at NRP. Again, Fred Worden professed to be impressed by my synthetic ability and writing skills and, "as a reward," decided to put me in charge of drafting the monthly Science Notes—a review of recent topics being bandied about in-house at NRP. This was a newsletter mailed out to all the Associates once a month. I took this as a small promotion, of sorts, though it was merely housekeeping.

Planning for a Momentous Meeting

The major focus of attention at NRP as winter turned to spring in 1977 was the upcoming Intensive Study Program (ISP) to be held, like its predecessors, in Boulder, Colorado, in July. The basic theme had been set a year earlier. It centered on the importance of local neuronal circuits and other phenomena localized to tiny regions of the brain. The subject matter ranged all the way from the strictly objective and well-known neuroanatomical facts of circuitry to highly speculative notions about unknown forms of information representation and undemonstrated mechanisms of information storage and consciousness. NRP's concern (that is to say, Frank Schmitt's concern) had become centered on two issues: (1) How to put together a roster of keynote speakers that did justice to that range of fact and speculation and (2) how to handle Gerry Edelman's new and presumably revolutionary but undisclosed theory of higher brain function.

My earliest encounter with these interrelated issues had come about at a science staff planning session in February. Schmitt had assembled us to report that Gerry Edelman had been doing some serious thinking about information in the brain and might be persuaded to give a talk on the subject on the last wrap-up day of the meeting. Schmitt further informed us that he was flying to New York the following day to discuss the matter with Edelman and another of NRP's senior Associates, Vernon Mountcastle. Upon his return, Schmitt reported that the three of them had agreed that Edelman would give a presentation of his theory on the final evening of the meeting, to be introduced by Mountcastle.

The notion that the final day of the ISP could be devoted to theoretical speculation had come as a surprise, since many NRP Associates, including Mountcastle, had debunked the idea of any attempt at an overall synthesis of the proceedings. Apparently having Edelman expound on his personal speculations about higher brain function did not fall into the same disreputable category. For my part, I was OK with it since, in my view, synthesis and creativity were what NRP should be promoting anyway. However, perceiving there to be a difficulty in scheduling two heavyweights like Edelman and Mountcastle on the same evening program, I asked if the two of them necessarily had to go together.

"Absolutely!" was Schmitt's immediate and emphatic reply.

I hesitated for a moment, then against my better judgment, went ahead and said, "Why?"

Schmitt recoiled as if bitten by a snake. "Mountcastle is a man of status!" he blurted. "Edelman is no neurophysiologist. He needs Mountcastle!"

Another issue was whether a third speaker might even be in order. In response to Schmitt's request for a suggestion, I said, "This would be a good spot for Ross Adey."

Schmitt burst out with a hearty laugh. "Come on, man, you must be kidding!"

I was informed in no uncertain terms that Edelman and Adey did not get along, that such a combination would sabotage the meeting, and that therefore it was totally unthinkable.

We quickly turned to the question of who would replace Mountcastle as a keynoter on the first day, if Mountcastle had to move to the end of the meeting. To my surprise, Schmitt himself suggested that Adey might be a long shot for such a role. He noted correctly that Adey had been at the forefront of many of the ideas that NRP intended to make a focus of the ISP—especially the importance of electrotonic activity in the extracellular space as a medium for information flow in the brain. But then he went on to note that Fred Worden strongly opposed Adey's inclusion on grounds that what Adey would have to say could not be predicted. (True enough, probably, but all the more reason to include him.) In time, I came to suspect that Worden's consistent opposition to Adey had more to do with their history together at UCLA than on what Adey would or would not say at any point in time.

Another factor surfaced as discussion proceeded. There appeared to be a certain lack of respect among, in Schmitt's words, "our group around here," for Adey's admittedly controversial work that often had been ahead of its time. Too much of it was "soft science" to the "group around here," by which was meant the circle of Associates to whom Schmitt was most attuned at that particular time. My argument that the ISP would profit from provocative rather than predictable keynote addresses went nowhere. In the end, Ross Adey, who more than anyone had pioneered the central theme of the upcoming ISP, was left out of the program entirely.

Palace Revolt

Though frequently rebuffed, we staff scientists began to assert our opinions more aggressively and display a growing indifference to the prevailing convention that vested all scientific judgment of any importance in Schmitt and the coterie of NRP Associates that he most looked up to. (At that particular time, they included, especially, Gerry Edelman, Vernon Mountcastle, and Maxwell Cowan—though the latter two themselves made no pretention to exceptional wisdom.)

Parvati Dev had been at NRP the longest and had the best rapport with Schmitt. She and the Program Director, Barry Smith, had coauthored the paper in *Science* on local circuit neurons that served as the leitmotif for the upcoming ISP. Bruce, Key, and I looked to Parvati and Barry as examples of what the in-house science staff at NRP could accomplish once fully engaged in a project. Thus, we began to study a variety of topics pending at NRP and to circulate commentaries and analyses on them. Schmitt had become interested in a topic dubbed central core control, having to do with pathways that emanate from the brain stem and use a particular family of neurotransmitters to broadcast generalized messages through broadly distributed higher brain regions. This topic was very close to Key's research experience, so he took the lead in its study and analysis. In early May, he distributed a thoughtful, thorough memo to all the science staff, soliciting feedback and opinions designed ultimately to lead to a planning session; and we responded in kind. This activity on our part had been encouraged by Fred Worden, who was genuinely grappling with a way to provide staff scientists with more professional satisfaction.

Schmitt was aware of this flurry of activity on the floor below and apparently became concerned that the topic was being taken from him. On a Monday morning in mid-May, we arrived to find that Schmitt had organized by phone the previous Friday evening an entire one-day conference, complete with eight or more committed speakers of his own choosing, totally without regard to all the groundwork that Key and the rest of us had done. To say that we were livid is an understatement. We demanded and got an immediate meeting with

Fred Worden, who had given his word that we would be incorporated more into the substantive scientific decisions at NRP. We railed not so much at Schmitt's arrogance in totally circumventing all our work but at his monumental discourtesy in doing so without so much as a mention to anyone or an acknowledgement that anyone but him had given the matter any thought at all. Worden sympathized with us—even admitted to being a little angry himself—but made the point that Schmitt was going to do whatever he wanted to regardless of our feelings.

The circumstances might well have hastened my departure from NRP had it not been for another assignment that gave me the opportunity to prove what I believed about the worth of Staff Scientists. Well over a year earlier, in November of 1975, a work session had been held at NRP on Specificity and Plasticity of Retinotectal Connections. The primary organizer had been the well-known and highly respected developmental neurobiologist Mac Edds, who had moved to NRP to become its Program Director. Gerald Schneider at MIT and R. Michael Gaze at the National Institute for Medical Research in England had agreed to cochair the meeting with Edds. Tragically, Mac Edds died suddenly just eleven days after the meeting had ended. This left responsibility for writing up the proceedings of the meeting for the *NRP Bulletin* in the hands of Gaze and Schneider. Gaze had submitted a preliminary draft of his section. Schneider had submitted nothing after sixteen months. Another work session was in danger of being lost from the printed record. At George Adelman's urging, Fred Worden agreed to let me attempt another resurrection. It would be, I hoped, a reprise of my 1968 salvage of the "Brain Cell Microenvironment," which in time had become one of the better known and more influential *NRP Bulletins*.

Hints of the Revolutionary Theory Revealed

Gone by the time the meeting began were all hopes on the part of Barry, Key, and me, at least, that there would be much creative integration arising out of the densely packed, detail-oriented program.

Inclusion of a few speakers with reputations for broad, restless minds—notably, Ted Bullock, Gordon Shepherd, and Jack Diamond—held out a little hope that a few telling insights or even rogue speculations about the implications of local neuronal circuits might bubble forth into the Rocky Mountain air by the time the two weeks were finished. Even if that didn't happen, the scheduling of Gerry Edelman for the entire final afternoon assured that the meeting would end with an exclamation mark of some sort. And contributing to *that* speculation in no small part was an aura of secrecy that had been wrapped around the substance of Edelman's presentation, primarily at his insistence but meekly indulged by Schmitt as well.

Gerry Edelman's long-awaited abstract for his presentation on the final day of the ISP arrived on May 3. Entitled "Group Degenerate Selection as a Mode of Higher Brain Function," it consisted of three paragraphs. The first two were as obscure as the title, but the third seemed to contain the heart of the matter: The fundamental assumption was that the brains of higher vertebrates function on selective principles, the unit of selection being a group of neurons whose basic pattern is defined during early development. We are born with or develop at an early age patterns of neural circuitry that represent the information that we will perceive, assort, and interrelate over the remainder of our lifetimes. Those patterns, which reoccur and are reinforced, become selected for persistence, thereby defining the way we see the world, or at least process information about it, forever after.

This was not a new idea. As previously discussed, the concept of cell assemblies as neural substrates for perceptual units had been popularized by Donald Hebb in *The Organization of Behavior* twenty-eight years earlier. The slightly new twist that Edelman was putting on it was the notion that only a fraction of the cell assemblies potentially present in the brain at birth are selected for persistence into maturity, and that selection is based on the actual postnatal experience of the individual. Doubtlessly, the concept was inspired by analogy with the immune system, in which a much larger genetic potential for antibody production exists at birth than will actually be selected for antibody production in later life, the selection being based on the particular antigens that the organism encounters. His work in immunology had won Edelman a Nobel Prize. In a sense, he was revisiting his previous triumph.

We scoured the abstract in vain for hints of a molecular mechanism that would constitute the selection process at the cellular level. If Edelman had a specific molecular mechanism in mind, he was evidently not going to disclose it in his abstract. There was no mistaking, though, that a monumental breakthrough was being hinted at; and Schmitt ate it up because it played so perfectly into his fervent desire that NRP really catalyze a new insight into higher brain function. Nothing less than the entire afternoon of the final day of the ISP would be devoted to a platform for Gerry Edelman and his new group degenerate selection theory of higher brain function.

Reflections on Running and Ending Marathons

In April, Penny came up from New York, and her brother, Johnny May, flew in from Houston for the Boston marathon. We loaded Johnny up with carbohydrates for two days, dropped him off in Hopkinton, then drove into Boston to the finish line at the Prudential Center. It was a very moving spectacle, watching thousands of runners end their twenty-six-mile ordeal in varying degrees of agony or ecstasy. Some staggered into the crowd helplessly just before the finish line; others collapsed a few yards from the end, only to get back up and limp across the line and collapse again, and some spurted in like they had just been out for a casual jog. There were young girls and old men, scrawny types and full-figured athletes, a great mixture of nationalities, and an audacious display of jogging wear and headgear. It was a spectacle unlike any I had ever seen—a glorious mixture of trauma, entertainment, and inspiration. I thought to myself what a great accomplishment it would be to run the Boston marathon just once, to be able to say you had done it. I doubted that I ever would, but the next morning, I got up at 6:00 a.m. and jogged for a third of a mile.

As Johnny was completing his race in Boston, a marathon of a different sort was nearing its inevitable conclusion in Memphis, to which the Ungars had moved when his grants were terminated and he lost his lab in Houston. Neurological symptoms had set in for Georges, indicating that cancer had spread to his brain. Two rounds of chemotherapy arrested the spread, but the third failed to do so. As spring turned to summer, he stopped eating and deteriorated quickly. When the end was clearly near, Alberte moved him home from the hospital, and he died in the early morning of July 26, 1977, with his wife and daughter at his bedside.

30

Beauty and Bluster in Boulder

I assume that everyone has one or two magical places that hold a special spell in a corner of their lives. Boulder is such a place for me. I first saw it on a golden September afternoon in 1962 on my first trip north and west of Texas. The gorgeous autumn aspens and cool, crisp air against the Rocky Mountain backdrop presented a scene of beauty unprecedented in my experience. Seven years later, Boulder became the first place I traveled to outside of Kansas upon receiving my PhD. In the late afternoon of July 20, 1969, I sat transfixed in the television lounge atop Darley Towers, a dormitory on the campus of the University of Colorado, to witness the fuzzy image of the first human to walk on the moon. For this combination of serious and trivial reasons, Boulder is no ordinary place for me. Thus it was that as Carol, Anthony, and I landed in Denver and drove the short distance to Boulder on the afternoon of June 16, 1977, the tinge of excitement that I felt at returning to a special place for a time overtook my prevailing skepticism about the likely disappointing outcome of this Third Intensive Study Program of NRP.

Hope, Hype, and Atmospherics

Our first staff meeting was held the following day. In what had become a pattern, the administration had billed it as "the most important meeting of the whole two weeks." This one consisted of Frank Schmitt and Fred Worden reiterating things we had heard several times before, followed by a boring hour or two in which Schmitt fiddled with the slide projector in the auditorium where he would give his opening address while we on the science staff sat in the audience, reading pulp paperback novels or whatever other reading material had been left behind by disinterested summer school students.

The weekend wore on in that vein—a mixture of boredom at NRP functions and excitement over the ambiance of Boulder. Following the opening reception on Sunday evening, which Carol and Anthony had skipped in favor of a trip to McDonald's, I joined several members of the NRP staff and one of NRP's most vaunted Associates for dinner at a Chinese restaurant in downtown Boulder. By that point in time, I had worked at NRP for six months and become quite obsessed with the question of whether NRP really had in fact influenced the development of neuroscience to a significant degree. I was surprised to hear the Associate offer the opinion that NRP's impact had been rather limited—"Less than they would like to believe." When Bruce Brandt pressed him on the question of whether NRP really played a catalytic role in creative synthesis and conceptual thinking, he observed that such attempts usually become preoccupied with organization and mechanics. Most of the creative people he knew, on the other hand, were rather disorganized. "They tend to have their most creative thoughts in the bathtub, or somewhere like that."

Keynotes and Coffee Breaks

The following morning was devoted to the keynote addresses of the Third Intensive Study Program of NRP, now after months of planning, finally underway. Maxwell Cowan gave the best talk—an excellent overview of developmental neurobiology. Frank Schmitt was second best, with an eclectic compilation of facts and issues reasonably well presented. Ted Bullock spoke about neural systems at different levels in different configurations. As usual, the substance of his message was stimulating, but his presentation was not. He gave his talk in a fumbling monotone, as though he had just composed it on the plane to Denver, which he may well have done since an NRP secretary reported that he handed her fourteen pages of handwritten manuscript for typing upon his arrival. The fourth keynote speaker was Vernon Mountcastle, whose pioneering work in cortical neurophysiology as much as anyone else's had laid the groundwork for the concept of local neuronal circuits. The power of the concept and the relevance of his discoveries to it were evident in his presentation for those who remained awake through his distinctly uninspired delivery.

The afternoon's program by contrast was more spirited. Keir Pearson, Gordon Shepherd, and John Dowling gave three excellent talks on the importance of information conveyed by graded potentials over local circuits within the limits of a single neuron. Their focus on the trees had given more clarity than the keynote speakers in the morning had given to the forest.

During the coffee break, I struck up a conversation with Ross Adey, noting that I had been surprised at his omission from the program of the ISP. He confessed that as plans for the program had progressed without him, he had become angry and had written in protest to Fred Worden, who had responded in essence that no one at NRP found his ideas particularly interesting and that he should "just get lost" (Adey's words). I asked him why Worden felt that way, and he said he didn't know. Adey asked me how I was doing, and when I told him about my grant and intention to pursue work on glycoproteins and gangliosides at the Shriver Center, he seemed genuinely pleased and interested. At several points in my career, Ross Adey's apparent interest in me and my work had provided a real if temporary ego boost, and this was one such occasion.

Back at the dormitories at the end of the day, dinner was pleasant, followed by long, lingering talks with various friends on the steps outside the dining hall in the golden afternoon sunshine that gradually dipped below the front range of the Rockies. Thankfully, there was no staff meeting to attend that evening, so the first day ended in beauty instead of bluster.

More Caffeinated Conversations

The first week of the ISP rolled onward—a dizzy blur of electrophysiological and neuroanatomical detail more than sufficient to dull the senses. The large plenary sessions were not conducive to interactive argumentation, speculative discussion, or the generation of creative insights. Deprived of this or other forms of intellectual stimulation, I mostly stirred restlessly in my seat and looked with longing toward the mountain valleys to the west. During particularly boring sessions, I took to reading real estate listings of land offerings in the area. The evenings were better, when informal discussions around dinnertime and afterward allowed us to ask the questions and kick about the ideas for which the formal plenary sessions made no allowance.

There were occasional bright spots during the day, usually around the coffee breaks. Ted Bullock and I had a stimulating talk about higher brain function. I asked him if he thought that complex cognitive functions or concepts require emergent properties of massive numbers of neurons, and he said, "Not necessarily." In his view, discreet and rather limited circuits could uniquely represent complex and abstract information. We both lamented the fact that it had been so difficult for neuroscientists to figure out the language of the brain's cells in representing information. He went on to say that this ISP was a good example of how NRP was failing to shed much light on that issue: every session was crammed with speakers, every minute accounted for, with nothing left to chance—the opportunity and incentive for spontaneity being thereby diminished. He said that he had urged Fred Worden and Frank Schmitt repeatedly to experiment with format, such as controlled confrontations between scientists with discordant views, and to encourage Staff Scientists and others to come up with bright ideas and hot new issues. I pointed out that we tried, but that our ideas got filed away somewhere and that was the end of them.

At another coffee break, I talked to Earl Stadtman, a biochemist from the National Institutes of Health who had gamely been sitting through the endless procession of brain potentials and electronic squiggles across our visual fields (if not our conscious awareness). We talked about how interdisciplinary innovation is brought about. Without discrediting the effort being mounted by NRP, he expressed the view that a neuroscience research program could not be built on the reputation of eminent scientists only, because neuroscience as we know it at one point in time will evolve into something different in the future, and those who are part of today's success will not necessarily be best equipped to deal with tomorrow's issues.

Disappointment and Creativity at the End of the Week

On Friday, the last day of the first week, I sat next to Ross Adey on the bus ride into the lecture hall. It was only then that I learned he was leaving that day. We talked about how we might generate interest within NRP for work sessions on gangliosides, the extracellular space, or like topics. When I suggested that the idea would carry more weight with the NRP hierarchy coming from him than from me, he laughed sarcastically and said, "That's the way it is in everything, isn't it? You have to reach the right threshold."

He had planned to show a few slides and make a brief commentary at the end of the morning's session chaired by Rodolfo Llinas, but Llinas didn't call on him soon enough, and he had to leave to catch his plane. All week he had sat at the back of the auditorium, refraining from asking questions or offering opinions—I assumed out of wounded pride and frustration. This proud man who, years earlier, had the vision to point out the importance of the topics now embraced by so many had been reduced by his professional rivals to that of a silent observer—barred from the program except for a single brief commentary. In the end, even that had been denied him.

The afternoon program was so bad that even Schmitt stood up at one point to deliver a brief haranguing and typically rambling lecture about the main points that each of the speakers was supposed to be making but wasn't. Ironically, one of the few NRP Associates who could have pulled some main messages out of the crush of information that had dulled all our senses was at that moment on a plane back to southern California.

Feeling a great need to get away for the weekend, we procured a car on Saturday morning from Colorado Rent-a-Heap and drove up into the Rockies, then back down to Denver where we spent the night with relatives. The following day, we basked in the swimming pool at their apartment complex, wondering how we might ever manage to move to that part of the country. At day's end, we drove back to Boulder and decided to take in a movie. In deference to our six-year-old son and in light of rave reviews, we went to see a newly released science-fiction thriller called *Star Wars*. In that way, our week finally ended on a creative note after all.

31

The Gerry Show

The second week of the ISP was easier for me to stay awake through since it was much more biochemically oriented. Monday, in particular, was stimulating because it dealt with neural development and because it was chaired by Gerry Edelman. One of the more impressive presentations was by Urs Rutishauser, a student of Edelman's that I had met previously at NRP, who was at that time in the early stages of research on NCAM (neural cell adhesion molecule)—a glycoprotein with a high sialic acid content. This research paralleled my own interest in the role of glycoproteins and glycolipids in the formation of specific neural connections. It differed from my work in that the effects of NCAM are very nonspecific, hence unlikely to provide the unique information needed for establishment of precise neural connections.

Again, there was much talk of neurospecificity based on selective stabilization of some synaptic connections at the expense of others. Jean-Pierre Changeux threw out the notion as though it were a new idea, and Edelman reinforced it with a tantalizing reference to the fact that on Friday, he would offer a new selectional rather than instructional theory of brain function. As the week wore on, excitement within the upper echelons of NRP rose in proportion to Edelman's insistence that the substance of his presentation be kept carefully under wraps. Frank Schmitt and Fred Worden took it all very seriously. Most of the younger scientists, and many of the veterans, thought it was a little silly. I was among a minority who simply felt frustrated. The fact was that a person of Edelman's brilliance would at the very least be articulate—and quite possibly, he would, in fact, be able to propound a radical new insight into a problem that had confounded lesser minds for generations. Anyone who already had won a Noble Prize for elucidating the structure of the antibody molecule deserved to be taken seriously. But to me, this was all the more reason for the written version of his presentation to be released beforehand, so it could be studied and analyzed in anticipation of the presentation itself, to better inform questions and commentaries that rightfully deserved to be aired. The decision was final, however: the manuscript would not be released before high noon on Thursday.

An Unanticipated Undercard

In the meantime, another storm was brewing that turned out to provide one of the highlights of the entire two weeks for me. Characteristically, by the time the ISP had begun, with its emphasis on local neuronal circuits, Frank Schmitt's attention had shifted to another breakthrough concept that he was sure would revolutionize understanding of brain function. That was the notorious "central core control" topic that Schmitt had snatched from the Staff Scientists back in May. On Wednesday evening after dinner, Key and I were dispatched to a meeting intended by Schmitt to be a simple ratification of plans for an upcoming work session on what by then was being called central core regulatory systems (CCRS).

A star-studded cast gathered in the basement conference room of Darley Towers after dinner. Present were Schmitt, of course, and several NRP luminaries: Vernon Mountcastle, Leslie Iverson, Robert Moore, Floyd Bloom, and Sanford Palay. A strong supporting cast was headed by Victoria Chan-Palay and Tomas Hökfelt. Bit players included George Adelman, Key Dismukes, and myself. Iverson, Moore, and Bloom—all three—had played major roles historically in the discovery of the importance of neurotransmitter-specific pathways in the brain (that is, circuits that are anatomically distinguishable and make use of a particular family of neurotransmitters). Palay was one of the most respected neuroanatomists in the world. Vicki Chan-Palay was his wife, a brilliant, energetic young scientist whose star was rising rapidly at NRP on her own merits.

Robert Moore began by reviewing the "enthusiastic" work session that Schmitt had put together singlehandedly back in Boston. Vicki followed with strong support for a full-fledged work session, noting correctly that Moore's suggested outline was lacking in certain physiological and behavioral topics. The discussion rocked back and forth over details among the advocates, all of whom seemed to be enthused about the topic. Through it all, Floyd Bloom had sat silent. When his opinion was asked, he drew a measured breath, and in a deliberate, almost apologetic tone, he proceeded to dismantle the argument for a work session piece by piece. The "central core regulatory system" concept was now more than a decade old, he pointed out, and had been treated thoroughly by NRP three years earlier. While there were new findings, adding some anatomical detail, "morphology doesn't prove function," he said, and no radically new conceptual picture had emerged to justify a meeting. In his view, a work session at this time would just turn into another rehash of the neurochemistry of the biogenic amines. For these reasons, he was opposed to it.

Jaws dropped all around the table, and Frank Schmitt looked like he had been clobbered by a sledgehammer. Didn't Floyd think, Schmitt asked incredulously, that these new types of neurotransmitters represented an important new finding? Everything on Moore's list was old hat, Floyd responded. The notion that some neurotransmitters might be modulators of passing neural impulses rather than direct signals between adjacent nerve cells (as for a classical neurotransmitter) was not a new idea and, in fact, was not even demonstrated by currently available data. The morphological evidence that chemicals might be released all along the axon rather than at the nerve ending only was equivocal, and even if it turned out to be right, it would not be a revolutionary new concept. Frank Schmitt's pet new project was going down in flames.

There then ensued one of the most stimulating intellectual clashes I ever had the privilege to witness. Vicki was as taken aback as anyone around the table by Floyd's devastating commentary, but she rallied. Slowly at first, then with growing passion, she came back at him point for point. She challenged his opinion that there was nothing new, arguing vigorously that the classical views of synaptic transmission no longer were adequate to explain the observations. New chemical families, in new circuits, with novel ultrastructural features were being uncovered monthly, she argued. The topic was hot, the time was right, and NRP would miss a major opportunity to have an impact on neuroscience if it declined to hold a work session soon, in her view. But every point that Vicki made was countered by a negative response from Floyd. With an equally articulate but disarmingly calm tone that contrasted with Vicki's passionate style, Floyd held his ground with stubborn firmness.

Their dialogue flowed back and forth for what seemed like an hour or more, with little commentary from anyone else. Two stars had faced off: a venerable reigning champion with the wisdom of experience, and a brilliant, young newcomer with the passion of youth. Everyone else stood aside to watch them go at it, intellect to intellect. Others began to pitch in eventually, but not until it was obvious that the match would end in a draw, and that meant no decision for Schmitt. While I admired and liked Vicki Chan-Palay tremendously, I couldn't help enjoying the mortal wound that Floyd Bloom had inflicted on Schmitt's latest brainchild. The problems that had surfaced that evening had been foreseen by Key and other staff scientists who had done the groundwork that Schmitt had chosen to ignore. Schmitt had wanted the show to himself, so now he deserved credit for its failure.

The next day in a postmortem commentary, Sandy Palay viewed the events of the previous evening merely as an example of how one articulate negative could torpedo a room full of positive attitudes. The truth was that neither Iverson nor Mountcastle—the two NRP Associates with the stature to challenge Bloom—had chosen to do so; and the others—sensing a deep, personal dispute between Vicki and Floyd (though the arguments had never become personal)—had decided to stay out of the fight. Mountcastle's only commentary, in fact, was not a word, but a little act of rebellion that I much appreciated and admired. At one particularly tense point in the debate of the previous evening, he had leaned back in his chair and sailed a paper airplane across the table toward George Adelman. It started off in level flight, then nose-dived to the floor. Whether this had been intended as comic relief or a metaphor for the fate of Schmitt's pet project was unclear, but it was obviously one of those little eccentricities that Mountcastle felt his maturity and staid conservative image entitled him to.

Advanced Notice at High Noon

Precisely at midday on Thursday, as participants filed out of the auditorium from the morning's session, Fred Samson, Bruce Brandt, Key Dismukes, and I stood at the doorway handing out the manuscript of Edelman's presentation. We would have twenty-four hours to prepare for what Schmitt had begun to refer to in historically monumental terms.

Leslie Iverson was in charge of the program on neurochemistry for this day. As one of the NRP Associates least interested in integration, he brought the afternoon session to a close, not with any kind of overview, but with yet more data from his own lab. And believe it or not, another session on neurochemistry was scheduled for after dinner; but I was sick of data and really wanted to study the Edelman manuscript, so I sought out a neighborhood MacDonald's, which could keep me in coffee for four hours while I studied through the details of what the morrow was supposed to bring.

I pored over page after page in tedious detail. It appeared that Edelman was going to offer a multifaceted, convoluted theory of neural information representation and consciousness. His idea of information representation was no more than Hebb's "cell assembly" concept, but his theory about consciousness did appear to offer some novel features. And it looked like he would propose an ontogenetic (developmental) theory for the persistence of some cell assemblies over others—that is, certain neural traces acquire meaning through experience, while others not similarly reinforced are lost. To make even this much sense out of the manuscript really did require four hours, and by the time I returned to our room and got to bed, it was well after midnight. But I was so wound up, it was another hour or so before I could get to sleep.

Main Event

Finally, the afternoon of Friday, July 1, 1977, arrived, and there was palpable excitement in the auditorium as participants and even quite a few spouses filed in after lunch to hear what NRP's most gifted Associate had to reveal. Curiously, following the program that he had chaired four days earlier, Gerry Edelman had kept a low profile—adding, of course, to the aura of drama surrounding the event that now was to unfold. After a brief and modest introduction, the months of mystery, intrigue, and hype culminated with Edelman at last standing alone before a tantalized but potentially antagonistic crowd of Nobel laureates, junior scientists, secretaries, and spouses. Uncharacteristically dressed down for the occasion in sports coat and open collar, he took the rostrum and assumed his usual commanding presence.

At the outset, he described his topic as "fascinating but disreputable" and went on to point out unconvincingly that he had tried to enlist "senior scientists more qualified than [him]" to tackle the issue without success. Then he plunged into the heart of his theory. Neural information is represented in functional groups of local neuronal circuits. The building blocks of experience are laid down by the time of birth or during early development in a "group degenerate" fashion, meaning that an array of similar but not identical circuits can represent the same information. Experience selects and sharpens certain circuits, which become perceptual units when they undergo a phasic reentry process, making them accessible to consciousness. There was much emphasis on consciousness, and the reentry mechanism was, for me, the most difficult part of the theory to grasp. It also appeared to be the most speculative.

For eighty minutes, he held—it is fair to say—the audience in rapt attention. His exposition was not solely scientific—he referred to philosophical concepts where appropriate, drew upon numerous literary allusions, and punctuated the tedious dialogue with a few episodes of comic relief. It was a true tour-de-force. Following the argument carefully for that long was a strain, but I managed to stay alert throughout. My study of the previous night had helped immensely. When Edelman finally ended, whatever skepticism remained in the audience—and private comments later revealed there was some—was buried by a sustained round of applause.

A number of people had questions in the extended question-and-answer session that followed. Each one gave Edelman the opportunity to expand and comment on his own theory. As usual, his handling of questions was masterful. Toward the end of the session, I managed to be recognized by Vernon Mountcastle, who was chairing the entire session and who called on me by name, to my flattered surprise. I asked Edelman (1) how the representation of information in his scheme differed from Hebb's cell assembly, (2) whether the changes required for storing information could occur somewhere other than the synapse, and (3) whether the site of neural computation could be subcellular, as on local arrays of dendritic membranes. To the first, Edelman confessed that he had only recently read Hebb, and his answer indicated that he was not fully acquainted with the cell-assembly concept. To the other two, he admitted that nonsynaptic sites could be the locus for plasticity and computation, but that he believed the synapse to be the most likely site for the types of changes envisioned by his theory.

By prearrangement, there were some designated commentators who dealt rather more gently with the theory than they need have. John Szentagothai's comments were probably the best. He wondered if the theory were yet mature enough to be useful, and Edelman responded that ideas can only be shown to be mature in retrospect. When Szentagothai used the double helix of Watson and Crick as an example of a breakthrough that occurred only when sufficient data were amassed, Edelman countered that the idea for the double helical structure of DNA resulted from Watson and Crick's "passion for a particular molecular structure," rather than the accumulation of critical data. I was impressed by this exchange. Whether or not Edelman was right, the notion that the impact of an idea is often difficult to judge in the present seems obviously to be true.

Frank Schmitt ended the session with the predictably hyperbolic statement that "this has been an historic day for neuroscience, possibly all of science." Then it was over. We filed out of the auditorium, a little bleary and certainly stimulated. But I hardly felt that we had witnessed a particularly historic event. What we *had* witnessed was a brilliant intellect conveyed through some masterful showmanship. But to his credit, Edelman, an immunologist, had stood up in front of the best minds that neuroscience could muster and propounded a theory in some detail that no one else had attempted. It had some truly insightful components and was, in many respects, credible. Only a person with Edelman's monumental ego would have tried it. As one who lamented so often the lack of boldness or creativity at NRP, I had to respect him for what he had done.

Time would show that, in the short term, Edelman's theory had no great impact in neuroscience. Articulate elaborations of it in subsequent versions over a number of years made it clearer and a little more persuasive. In the final years of his life (Edelman died in 2014), he brought what I came to appreciate as some genuinely useful insights to the scientific study of consciousness.

Aftermath

The final Friday night of the ISP was given over to a banquet, served with generous rounds of wine that seemed to facilitate the flow of more exaggerations from the head table about NRP's role as the cutting edge of neuroscience. During the speeches, it was disclosed that July 1 was Gerry Edelman's birthday, so to his considerable embarrassment, some of the more boisterous diners sang him "Happy Birthday." Jack Diamond—the man who had led us all into inebriation on the snowbound evening in Ontario the previous fall—noted Edelman's embarrassment and added to it by asking in a booming voice that the whole hall could hear, "Why are you humble now, after being so bloody arrogant all afternoon?" There was a roar of laughter, with enough sharpness in it to betray a perceived truth in the question.

After the banquet, Carol and I adjourned with a number of others to the Dark Horse, a pub across the way. Jack Diamond asked Carol to dance, so they partied for half an hour or more while I sat outside talking to Fred Worden's secretary about his inability to be an effective leader in Frank Schmitt's shadow. After a time, the party wound down, and we retired to our rooms. Carol was feeling great because (1) Jack Diamond had correctly pronounced her the best-looking woman in the entire NRP entourage, and (2) she had gotten to shake the hand of Jean-Pierre Changeaux, a charming Frenchman and Nobel laureate. On the other hand, I was depressed. It was partly fatigue and partly the recurring feeling—always stirred up by exposure to very accomplished scientists that the ISP had provided in abundance—that my career kept falling short of its potential. Too restless to go to bed, I went out for a walk, looking for someone to talk to. But there was no one. The ISP was over, and except for the stimulation of a couple of hours on the final afternoon, it had been about as unsatisfactory as I had predicted. It was July 1, 1977, and my career as a neuroscientist was mired in a dreadful calm.

32

Haystack Full of Needles

In the years that followed the death of Georges Ungar, some remarkable advances would be made in understanding the cellular basis of neural plasticity. And neuroscience would continue to expand in scope and break into ever more specific subdisciplines. It would become institutionalized in academic departments, as it already had at the University of California at Irvine and the University of Texas Medical School in Houston. But the growth and institutionalization brought little added coherence to the field. The scientific triumphs of neuroscience would march forward, but the romantic age of neuroscience, when it was possible to propose comprehensive if inadequate explanations of neural function, was passing away.

That neuroscience as I had known it was dead did not occur to me on the day I learned of Ungar's death, nor on the day after. Neither did I recognize that in the place of romantic neuroscience would arise a myriad of scientifically rigorous, vital, and productive subdisciplines that would add vastly to our knowledge of the nervous system in the ensuing years. The realization that a new day would come in neuroscience did not dawn on me until a month later; while relaxing in the White Mountains of New Hampshire, I would finally admit what most critics of behavioral neurochemistry had been saying for a long time.

Molecules and Memory on the Shores of Lake Winnipesaukee

Robin and I were reunited for the first time since I had left Detroit on Sunday afternoon, August 7, at Logan Airport in Boston. This was the gathering point for a host of scientists that would be spending the week at the first Gordon Conference on "Macromolecules and Behavior." The Gordon Conferences were started by chemists who wanted to get away from their work routines and families for a week of uninterrupted scientific discussion, argumentation, and meditation. Over the years, the conferences had broadened in scope and developed into a favorite summertime diversion for scientists from all over the world. Typically they were held at a boarding school or resort in the White Mountains of New Hampshire, where the scenery and clean air were presumed to be conducive for creative thinking and relaxation. Brewster Academy at Wolfeboro, on the shores of Lake Winnipesaukee, had been selected for the first in a series of annual conferences on behavioral biochemistry.

It was a little strange that at this point in history, there would be a conference on this topic.

The death of Georges Ungar had left the biochemical transfer of behavior approach with no proponent of stature, and there were not very many other proponents of biochemical correlates of behavior in the intact animal still pursing that type of research as their major activity. Indeed, on the bus ride to Wolfeboro, Robin proclaimed, with more prescience than I appreciated at the time, that the purpose of this conference was to "see if behavioral neurochemistry is dead."

Victor Shashoua, on the other hand, with dogged persistence was continuing to claim that three protein fractions are specifically expressed when goldfish learn to swim upright after having a float attached to their bellies.[1] Shashoua was the organizer of the conference, and he had enlisted the participation of a good mix of pioneers—like Agranoff, Rosenzweig, and Bennett—and those who were developing alternative approaches, like Timothy Teyler and Gary Lynch with the hippocampal slice preparation, Bruce McEwen on brain-hormone interactions, and Stephen Arch on egg-laying hormones in sea snails.

The pioneers did little to persuade us that behavioral neurochemistry as it had been practiced still had a future. Mark Rosenzweig talked about total RNA changes during learning, as though nothing had happened since 1960. Willem Gispen and Adrian Dunn left the impression that just about every biochemical change that had ever been ascribed to behavior could equally be attributed to the effects of stress and hormones. Bernie Agranoff gave a rehash of his old data on goldfish, offering nothing new; but he did propose a revised version of his "needle in a haystack" analogy about finding memory molecules. If they exist, he noted, the problem had become one of "a haystack full of needles." Indeed, it had long been the practice for neurochemists interested in mechanisms of memory to attribute a role in memory formation to their favored molecule of the moment, as Shashoua's far-fetched model illustrated. A few, like Eric Kandel, could plausibly argue that they were in fact looking at the biochemical basis of neural plasticity.[2] As the protein synthesis inhibitor studies had shown convincingly, protein synthesis was indeed necessary for memory consolidation—but which proteins, and where, and how? Agranoff's observation that the enigma of memory storage was not lacking in candidate molecules or mechanisms was doubtlessly correct, but hardly a comforting insight, however accurate it might have been.

Letting Go

Through it all, I persisted in struggling to make the data that Robin and I had obtained from operantly conditioned pigeons seem important and capable of telling us something about mechanisms of information storage that we could not obtain any other way. At midweek, I was still arguing with Robin that we needed to finish analyzing all the biochemical and subcellular fractions that were in the freezer in Detroit—as many needles as we possibly could.

But I was also listening to the newcomers. David Wilson showed through two-dimensional electrophoresis that a single ganglion produces hundreds of polypeptides—a fact long assumed but now convincingly demonstrated. William Greenough, one of the best students to come out of the Krech, Bennett, Rosenzweig, and Diamond labs, showed more convincingly than his mentors had that structural changes in the dendritic arbor of some brain areas in differentially reared animals do indeed occur.[3] Of all the newcomers, though, Timothy Teyler and Gary Lynch fascinated me the most. Teyler[4] and Philip Schwartzkroin[5] had introduced the hippocampal slice preparation to the United States, and Lynch had used his eccentric personality to broadcast its merits to anyone who would listen.

The merits of the hippocampus itself were already well-known. It is the remnant of what used to be the greater part of the forebrain of ancestral reptiles. In modern mammals, however, it has become compressed and buried beneath the more expansive cerebral cortex. In rats, it forms a long horseshoe shape in the interior, middle of the brain. When cut in cross section, it looks like a jelly roll because the neuronal cell bodies all line up in regularly spaced layers. The original anatomists thought the cross sections looked like the side view of a seahorse, with the head curved back on the rest of the body—hence the Latin word *hippocampus*, for "seahorse." From the brilliant visualizations of Santiago Ramón y Cajal at the turn of the century, it was known that the hippocampus is a collection of nerve cells that make four synaptic relays, replicated in parallel circuits, redundantly down the length of the tissue. If the hippocampus is divided into a series of cross sections down its long axis, like slicing a loaf of bread, each slice will have a complete set of all four neurons.

The other fact that makes the hippocampus particularly attractive for students of learning is that it appears to be involved in the encoding of memory. Patients with a damaged hippocampus have very poor short-term memory. Rats with lesions of the hippocampus have great difficulty in learning new tasks, especially those requiring spatial orientation. It seemed at the time that the hippocampus provides the brain with an internal map of the external world and is the primary screen on which changes in the significance of particular points in the external world are first recorded.[6]

Lynch and Teyler, in turn, reviewed these features of the hippocampus, then described their research on how brief stimulations within this four-neuron circuit could result in alterations in the threshold for subsequent activation of the circuit.[7] This example of long-term potentiation (LTP) in a precisely defined neuronal circuit that could be studied in vitro mirrored discovery of LTP by Tim Bliss and Terje Lømo[8] in the intact rabbit, providing an undeniable example of plasticity and immediately suggesting a cellular mechanism for memory encoding in a mammalian brain.

A genius of the Gordon Conferences is that no meetings are scheduled after lunch, when the nature of most humans and their descendants outside of northern Europe is to sleep or play. The afternoons are left for relaxation, recreation, the informal conversations that are the lifeblood of scientific interchange, and whatever other activities individuals or groups feel compelled to engage in. For me, it was tennis or volleyball, followed by reading and napping at the water's edge. I don't remember whether it was during one of those restful interludes Wednesday afternoon or in the middle of another intense conversation over bourbon and beer that night, but at some point, something snapped at a long-frayed breaking point in my thinking about learning and memory. I suddenly saw the hopelessness of pursuing the search for biochemical correlates of learning in whole, behaving animals with the technology available at the time.

On Monday, Agranoff had gone through the motions of describing research that was decades old, as if only because he had been asked to do so. Adrian Dunn, in another conversation, had told me he had never been more pessimistic about this line of research; the gulf between questions and answers just seemed as great as it had ever been. No matter how sophisticated the experimental design, no matter how many subcellular fractions and time points, no matter the depth of biochemical detail, the whole brain in the intact behaving animal was going to be operating in too complex a way, with too many variables and totally inscrutable processes, for anyone to make a necessary and sufficient connection between a specific metabolic process and a cognitive event.

Not that the model systems could do it either. The hippocampal slices cut away from their inputs from and outputs to other parts of the brain hardly constituted a cognitive unit. But what they did provide was a definitive neural circuitry clearly tied to the mediation of specific neural processes. If scientists could at least agree on that, perhaps they could then move on to the more detailed electrophysiological, anatomical, and biochemical dissection of the functional unit. Then by building from the ground up, perhaps the mechanisms of information storage and processing at each integrative level, from individual molecules to complete neural systems, could gradually be assembled.

Given the years that I had invested in behavioral biochemistry, the percolation of these heretical thoughts into my consciousness was at first unwelcome. Like a boat that strains against its moorings as the tide pulls it out to open waters, the baggage of cherished ideas conceived in the library stacks at Texas Tech on the hot summer nights of 1965 and brought to life so vividly in Georges Ungar's lab a few months later proved finally too heavy to hold. At some point on Wednesday afternoon, or Wednesday night, or Thursday, I finally let go—to settle for something smaller but more attainable. Embedded somewhere in my brain was the sense that it is better to leave behind small accomplishments than grand illusions.

For the remainder of the year, I grieved to a degree over Georges Ungar's death. At first I told myself it was because I had lost a valued champion. But in time, I came to suspect it was also a feeling of guilt at abandoning the research that had been his passion. It appears to be one of the profound realities of life that it sometimes takes the death of our heroes to set us free.

[1]   Shashoua, V. E. 1976. Identification of specific changes in the pattern of brain protein synthesis after training. *Science* 193: 1264–6.
[2]   Brunelli, M., V. Castellucci, and E. R. Kandel. 1976. Synaptic facilitation and behavioral sensitization in Aplysia: possible role of serotonin and cyclic AMP. *Science* 194: 1178–81.
[3]   Floeter, M. K. and W. T. Greenough. 1979. Cerebellar plasticity: modification

of Purkinje cell structure by differential rearing in monkeys. *Science* 206: 227–9.

4   Teyler, T. J. 1976. Plasticity in the hippocampus: a model systems approach. *Adv Psychobiol* 3: 301–26; Alger, B. E. and T. J. Teyler. 1976. Long-term and short-term plasticity in the CA1, CA3, and dentate regions of the rat hippocampal slice. *Brain Res* 110: 463–80.

5   Schwartzkroin, P. A. 1975. Characteristics of CA1 neurons recorded intracellularly in the hippocampal in vitro slice preparation. *Brain Res* 85: 423–36.

6   The function of the hippocampus today is viewed in a more nuanced light. While spatial orientation remains a critical aspect of its function, not all the neurons in the hippocampus respond to location in space. The idea that the hippocampus ought to be viewed more as a memory map than a spatial map was advanced by M. L. Shapiro and H. Eichenbaum. 1999. Hippocampus as a memory map: synaptic plasticity and memory encoding by hippocampal neurons. *Hippocampus* 9: 365–84. Empirical support for this view has been published by K. Z. Tanaka, H. He, A. Tomar, K. Niisato, A. J. Y. Huang, and T. J. McHugh. 2018. The hippocampal engram maps experience but not place. *Science* 361: 392–397.

7   Teyler, T. J., B. E. Alger, T. Bergman, and K. Livingston. 1977. A comparison of long-term potentiation in the in vitro and in vivo hippocampal preparations. *Behav Biol* 19: 24–34; Landfield, P. W. and G. Lynch. 1977. Impaired monosynaptic potentiation in in vitro hippocampal slices from aged, memory-deficient rats. *J Gerontol* 32: 523–33.

8   Bliss, T. V. and T. Lømo. 1973. Long-lasting potentiation of synaptic transmission in the dentate area of the anaesthetized rabbit following stimulation of the perforant path. *J Physiol* 232: 331–56.

33

Confessions of a Behavioral Biochemist

I returned to Boston from New Hampshire enthused about getting my new research direction underway. My grant from NSF on biochemical correlates of retinotectal neurospecificity was already geared toward my newly acquired accommodation with more limited and specific objectives. Bob McCluer had paved the way for me to start work on the grant at the Eunice Kennedy Shriver Center in Waltham on August 1. A key to making the project work was being able to hire Carol as a research associate on the grant, as I would have to be tied up three days a week working at NRP. Fortunately, NSF agreed; so the first week in August, Carol returned to work, and we became an official research team for the first time in our professional careers.

It was wonderful to have Carol working in the lab, even though my particular research techniques were new to her. Compared to the students with whom I had become accustomed, she learned so much faster and was so much more skillful that we progressed with a speed outside my previous experience. All I had to do was show her a procedure or often just write out the protocol, and she would carry it out flawlessly and usually with suggestions for improvement. Her first task was to learn the extraction procedure for gangliosides. It was obvious that for biochemical dissection of tissue as small as embryonic optic tecta or rat hippocampal slices, we would have to scale down the procedure.

So we took a method that Bob McCluer had originally developed at Ohio State University and cut down the volumes of all the reagents and reactants (essentially the same as downsizing the ingredients in a recipe from a dozen cookies to one), and soon we were isolating gangliosides from as little as one milligram of dried tissue.[1]

Since I was already used to dissecting specific regions of the rat brain, obtaining good hippocampal slices was not particularly difficult. The real advantage of the hippocampus, though, was the uniform alignment of its cells, making possible the isolation of layers consisting primarily of cell bodies or dendritic fields, for instance. Falling back on techniques I had learned in graduate school, I fashioned some tiny glass needles and, with them, managed to cut out a minute chunk of almost pure molecular layer (dendritic and synaptic fields) from a hippocampal slice the first week in October. That small technical accomplishment meant more to me than all the pointless meetings and inflated rhetoric that I was subjected to during my first ten months at NRP. Monday through Wednesday I spent at the monastery in Brookline, stuck in my dark corner office, brooding about my ineffectiveness. But on Thursdays and Fridays, Carol and I would drive to the Shriver Center together, work hard in a lively lab with a view of the Boston skyline in the distance, and feel good about ourselves by the end of the week.

A Retirement, a Request, and a Nobel Prize

Every good idea I had or move I made in a research direction at the Shriver Center seemed to be counterbalanced by a setback at NRP. By the fall, it had become perfectly clear that Fred Worden and Frank Schmitt had no intention of giving the Staff Scientists any meaningful responsibilities. When it finally dawned on me that Key Dismukes, Bruce Brandt, and I were there for the sole purpose of decorating NRP with scientific respectability—without us, it looked too much like just a country club for elite neuroscientists—I resigned myself to speaking when spoken to, reacting in whatever trivial fashion the situation called for, and otherwise concentrating on my own projects. In this mode,

I thus made rapid progress on writing up some data left over from Detroit[2] and in starting a new project based on my chance notice in the library one day of a solicitation from *Perspectives in Biology and Medicine* for essays for their Second Writing Contest for Authors 35 Years Old or Younger. My revelations from the Gordon Conference would be the topic of my essay. I would try to get down on paper the thoughts that had been festering for years, and I would make it even perhaps a veiled testimonial to the biochemical transfer of behavior paradigm that, with the death of Ungar, would now fade into history.

As Fred Samson never tired of reminding me, however, it was "a privilege to sit in that room"—the conference room at NRP. To be sure, for those tough enough to take the bruises to their egos, mere exposure to the quality of scientists who appeared there and the stimulation of their ideas were an experience not attainable anywhere else. While Fred was tough enough to take the bruises, I wasn't sure that I was. Nonetheless, I stayed because of the element of truth in what he said. I also stayed because I had agreed to, and didn't have another source of income anyway.

With this mixed set of motivations, I was looking forward to the Stated Meeting of NRP Associates in October, mainly because Ross Adey's term as an Associate had come to an end and the meeting would surely pay homage to his role as a creative force in the development of neuroscience. At the opening dinner on Sunday evening, Frank Schmitt was indeed gracious in his praise of Adey, and Adey responded in kind. I thought it all a little strange, in view of the way Worden and Schmitt had shut Adey completely out of the Intensive Study Program that summer. But Adey gave no indication of resentment. Perhaps he would have something to say in his farewell address on Tuesday. If he did though, it was too subtle for me to catch. Again he paid homage to the great influence that NRP had had on his career and to the unique advantages of the organization, but he went on to warn that NRP was in danger of losing some of the very things that had made it so good in the beginning. Since his term was finished, however, he didn't feel it proper to suggest the correctives; he merely hoped that the remaining Associates would give it some careful thought. The speech was elliptical and obscure. I was tempted to interpret is as a skillfully executed double entendre—on the surface, an almost obsequious praise of NRP, while underneath, attacking what NRP had become. But to what was Adey referring? The growing conservatism of NRP? Its excessive reliance on a small inner circle? Its loss of daring to plunge into new or untested waters? Perhaps none of these. Perhaps I read more into his remarks than he intended.

There was no reason for anyone to be left wondering. Forty-five minutes had been allotted for his presentation, and he had taken twelve. Schmitt's only remark after the applause had died down was "I'd like a copy of that." Worden's reaction as chair of the session was worse. Instead of using the remaining time to draw out the discussion that Adey's provocative remarks had invited, he simply stated with characteristic blitheness that a committee had been formed to look into the future of NRP. Both Schmitt and Worden had either missed the point of Adey's remarks or chosen to ignore it. Unfortunately, this was not the last time I would see a valued Associate go out with a whimper instead of the bang they deserved.

During a coffee break earlier in the day, Ted Bullock had cornered me in the hall and proceeded to give me a lecture on how Staff Scientists should take more initiative in developing projects at NRP. As an example, he raised again the possibility of friendly adversarial debates, like between himself and E. Roy John. I told him I couldn't agree more, but that all our initiatives were routinely either batted down or ignored. Like Adey, he responded with chagrin.

The Francis O. Schmitt Prize in Neuroscience for 1977 was awarded to Roger Guillemin for his chemical characterization of thyrotropin releasing hormone and other hypothalamic releasing factors. I first learned of Guillemin's work while at Baylor in 1966. Later as a graduate student, I had written him with some questions about his research, which he kindly had answered. He, too, was attending the October Stated Meeting of Associates at NRP, preparatory to giving his acceptance lecture for the Schmitt Prize. In the wake of Ungar's recent death, knowing of Guillemin's long friendship with Georges, I asked him what he thought of Ungar's influence overall. He answered solemnly, with a measure of sadness. In some areas, he said Ungar's influence had been very important and was recognized as such—the work on antihistamines, for example. In other areas, like his work on the biochemical transfer of morphine tolerance, his work had been far ahead of its time, and its importance had been underrated. In the case of his work on behavioral transfer, Guillemin felt that the research was dubious and very hard to accept, but it was the reliability of the work itself—especially the difficulty of the bioassays—rather than the antithetical nature of the concept, that was to blame. He ended by reiterating his great fondness for Georges and lamented the sadness of his passing.

I went on to thank Guillemin for responding to my query about releasing factors with a personal letter back when I was a lowly graduate student, and I told him I thought he was very deserving of the Schmitt prize. He patted me on the shoulder and told me how kind of me it was to say so. Nine days later on the way to work, I heard the news that he had been awarded the Nobel Prize for Physiology and Medicine. In science, sometimes nice guys finish first.

Back to Work in the Corner in the Dark

Back at work in my dungeon, I had decided the remainder of the year would be devoted to writing my essay on behavioral biochemistry. My introspections on this subject had begun during my years in Detroit and intensified during my forced detour into conceptual research over the past few months. At the Gordon Conference in August, my ties to past approaches had snapped, as I finally admitted that even the simplest behavior in animals as complex as vertebrates could not be understood in molecular terms by current technology. We would have to scale down our level of analysis, from the whole brain in the intact animal to a more manageable unit of operation, like the retina or olfactory bulb or hippocampus. Our objectives would be more limited, but our hope of making progress would be more realistic. This was hardly an original idea of mine; Eric Kandel, Tom Carew, Stephen Arch, Tim

Teyler, and many others were building entire careers on it. But I would make the argument as a reluctant convert to the ideology of lowered expectations—perhaps a perspective novel enough to garner attention.

At an emotional level, a key issue was whether the fifteen years that I and others had invested in the former approach had been a waste. On this point, I would argue not so. Overly ambitious and naive, yes, but not devoid of useful insights and the accumulation of a little bit of wisdom. It would be a wistful reminiscence of a noble effort by scientists who thought large and risked a lot. It would, in other words, be self-justifying.

Composing the essay made me think harder and deeper about these issues than ever before. It was at once the most difficult and elating project I had attempted since my doctoral dissertation. By late November, I had completed a rough draft, which I distributed for comment to George Adelman, Key Dismukes, Fred Worden, and Frank Schmitt. George was highly enthusiastic and not too critical. Key thought it was very interesting but needed for me to identify my intended audience more precisely and give added experimental details. Worden likewise found it interesting but too negative. Schmitt took the longest to respond, and when he did, his minimal comments were sketchy and indicative of the fact that he didn't really understand the point. Nonetheless, I incorporated his suggestions and those of everyone else in the revised version. Carol also read it and had comments similar to those of Key. His detailed critique and her suggestions were most helpful and made the final document a much better product.

The heart and soul of my research philosophy for over a decade; my tribute to Georges Ungar; my salute to the pioneering efforts of Bennett, Rosenzweig, Agranoff, and Barondes; my sympathy with the latter-day realism of Dunn and Gispen; and the confession of my own recent conversion were all distilled into a manila envelope and dispatched to the University of Chicago Press on December 16, 1977, the last workday of my first year in Boston.

Blizzard of '78

On January 20, 1978, nearly two feet of snow fell on Boston. It was the biggest snowstorm I had ever seen, and I felt sure I would be long gone from New England before I would ever see its equal. I was wrong.

Monday morning, February 6, dawned bright but overcast. A low-pressure area was moving up the Atlantic coast, developing some strength. The possibility of substantial snowfall by dark was predicted. About midday, the snow began to fall, gently at first, in a soft breeze. It being a Monday, I was working at NRP, and Carol was at the Shriver Center. As the afternoon wore on, the wind picked up with seeming malevolence, whistling though the creaking windowsills of the Brandegee Estate, plastering snow against the panes with alarming force. The management encouraged us to go home. Key and I accompanied each other to Newton Centre, where he lived. The roads were treacherous and rapidly becoming impassable. We barely managed to push his car off the street into his driveway, the snow now driving into our clothes and freezing our skin. The Volkswagen Camper that Carol and I had bought in New York and brought from Detroit was a constant source of irritation because the engine never had power and something was always broken; but on this afternoon, I was grateful to be in it, for few other vehicles would have had the traction to get me from Newton Centre to our new house in Newtonville.

Now home alone, I watched and worried for Carol and Anthony. As dark was descending and the storm was assuming blizzard proportions, they pulled into the driveway. By good fortune never explained, Carol had managed to get from the Shriver Center to Anthony's day-care center, then from there through the back roads of Waltham and Newton, slithering and sliding from roadside to ditch but never becoming completely stuck. Tens of thousands of other commuters were not so lucky. Caught in the rush hour to get home, the driving snow slowed them, then stopped them, then buried them. Not until nightfall on Tuesday did the wind die down and the snow cease to fall, and by then, it was too dark to see the devastation.

In the bright still sunshine of Wednesday morning, the storm's toll could first be gauged in its full and terrible extent. Tens of thousands of cars were stranded on Route 128 alone, a snake of vehicles thirty miles long flash-frozen at rush hour. Miraculously, deaths were limited to barely a score, but a miserable and frightening night was spent by thousands in the prisons of their buried automobiles. Governor Michael Dukakis declared a state of emergency on Monday night that lasted all week, banning all but essential vehicles from the roads and keeping everyone not in a critical occupation at home. I didn't have to call the governor's office to see if conceptual neuroscience was a critical occupation, but I did regret the setback to our experimentation schedule at the Shriver Center. There would be blizzards again in New England and destructive tides wreaking havoc on the shores north and south of Boston in later years, but no one who lived through the blizzard of '78 would ever admit to a greater winter ordeal.

Storm of Another Sort

Through the vicissitudes of my second winter in New England, I persisted in my effort to make the Staff Scientist position at NRP count for something. With my essay on behavioral biochemistry out of the way, the only meaningful project I had left was resurrection of the ill-fated work session on retinotectal neurospecificity. Following the untimely death of Mac Edds just days after the work session ended in November of 1975, Michael Gaze and Gerry Schneider, the remaining cochairs, were commissioned to write up the proceedings of the meeting for the *NRP Bulletin*. Gaze had indeed submitted a write-up for his portion of the program, but Schneider, who was responsible for only five speakers, had produced nothing. There the project languished until George Adelman proposed that I try to do something with it. George knew I could do it because I had done it before. Fred Worden was dubious, but didn't stop me from trying.

It had become obvious that Gaze's highly selective write-up was not going to satisfy a number of the participants, so I ended up proposing to do a rough draft of the entire meeting. Reconstructing the original presentations, capturing the nuances of the discussions, integrating the figures into a consistent format, and managing to reflect the considerable and often acrimonious differences among the participants had turned out to be no easy chore. But by February of 1978, my first draft was completed and in the mail to all the participants for their review.

The bright Wednesday morning after the blizzard of '78 was just the calm before a storm of another sort. Initial reaction to my write-up, now entitled "Plasticity in Retinotectal Connections," had ranged from lukewarm to enthusiastic. A number of participants doubtlessly were just happy to see that a manuscript might come out of the meeting at all. Others felt that my draft, subject to some editorial improvements, was a fair summary of the proceedings. On the afternoon of March 29, however, Marcus Jacobson called me to let me know in no uncertain terms that he found the draft report atrocious and totally unacceptable.

I had first seen Jacobson in action at the 1968 ISP where he had brilliantly and effectively confronted Paul Weiss, the virtual founder of the field of neurospecificity. Originally from South Africa, Jacobson had been a student of Gaze in England but, over the years, had fallen out with his mentor. Indeed, he had apparently fallen out with most everyone. His brilliance was questioned by no one. His methodology was questioned by some. His vitriolic personality was known to many. He was probably the single greatest source and cause of acrimony in the field of neurospecificity, though the field hardly lacked for villains. With another brilliant and acrid individual—Paul Weiss—as founder of the field, it perhaps could have been no other way. This was my day to experience Jacobson's antagonism firsthand. The product submitted for his review not only was atrocious, but NRP, in his view, would be doing a disservice to the field to publish it. And if it were published, he would forbid his name to be associated with it in any way. Furthermore, I was assured, he spoke for the majority of the participants. I pointed out, probably in too meek a manner, that the manuscript was submitted for him to change. If he didn't like it, he was supposed to make it right. But he was not interested in wasting his time on a piece of work already so far from the possibility of salvation.

If the intent of Jacobson's diatribe was to rattle me or worse, it succeeded in at least the former. Suspicious of his allegation that he spoke for the majority though, I tried to get him to name names, and at first he refused. When pressed, he suggested that Ronald Meyer and Kevin Hunt were among the ones who were "quite upset." I decided to call Meyer and Hunt. After long conversations with each of them, I concluded that they were indeed very dissatisfied, but might be willing to help make the changes that would make the manuscript acceptable. I tried to convey the impression that I had no ax to grind in the field and was a person they could trust to be fair. I was counting on the possibility that being confronted by an unknown person outside any of the existing ideological camps in the field would be novel enough to put them off-balance long enough for me to try to gain their confidence. They agreed to think about it.

In the meantime, I received a call from Sansar Sharma at the New York Medical College. Sharma was another student of Gaze but had remained basically loyal to his mentor (and therefore out of favor with Jacobson). Sharma had grown close to Mac Edds by the time of the work session, had labored with him to put the program together, and was devastated when Edds had died. Now he was calling to urge me emphatically to see the project through. By what incredible coincidence his call after a month of silence came within two hours of my conversation with Jacobson, I do not know, but it had the great benefit of providing me with leverage to tell the recalcitrants that "some of the participants are adamant that a *Bulletin* on this work session be published." The unstated implication was that something was going to come out, so they could be a part of it or not, as they chose.

In the days that followed, more responses came in, most of them positive. Two weeks after my phone conversation with Ronald Meyer, a gratifying letter arrived from him saying that our talk had significantly quelled his concerns. All he needed was assurance that the editing would be fair and evenhanded. He enclosed a balanced, scholarly critique of Gaze's contribution, which I knew would be a tough sell, but thought if I could win Gaze over as well, the rest would be downhill. But in the meantime, I had forgotten about Gerry Schneider, mistaking his silence for indifference. As the months to come would show, his silence reflected pathological procrastination, but hardly indifference.

A Prize Not Worth Noting

Arriving at the Shriver Center on Thursday mornings, after struggling through the frustrations of the first three days of the week at NRP, was like taking a deep breath of fresh air. Carol and I were rapidly proving ourselves to be the best research bargain in the Boston area; the Shriver Center had picked up two experienced PhDs and the overhead income on a $40,000 grant from NSF for nothing more than the cost of a couple of benches in a corner lab.

While my ultimate interest in the hippocampus was its utility for studying functional plasticity, at that point, we found ourselves working at a mental retardation research center and funded by a grant to study developmental changes in the nervous system; so we thought it best to pursue developmental changes in the hippocampus as well. The hippocampus in the rat is well suited for developmental studies because it matures in a gradient fashion (roughly from its central to peripheral borders). By careful dissection, different parts of the hippocampus can yield· a collection of tissue at different developmental stages in the same animal. This allows the experimenter to analyze for differences related strictly to the stage of maturation in the nervous system, apart from changes related to the age of the animal. By carrying out these types of studies, Carol and I showed that the ganglioside pattern becomes abruptly more complex at about the time that immature brain cells stop dividing and start sending out dendritic and axonal extensions. In particular, the larger ganglioside molecules—the ones with more sialic acid residues—become more prominent once the brain cell begins to differentiate into its final form.[3] Not coincidentally, it is now known that the complex gangliosides tend to disappear when brain cells regress to their immature state and start multiplying again, as when they form tumors.

On Thursday morning, March 2, I was upstairs in the animal quarters suffering through another setback in a recent series (a rat presumed to be pregnant was not), when our lab colleague, Peter Daniels, took a phone message from a Dr. Landau in Chicago. The message was conveyed to me as he had received it, brief and without details.

"He gave his name as Richard Landau."

While I was searching my memory for a hint of who Richard Landau might be, Peter continued with something like, "He says he's the editor of *Perspectives in Biology and Medicine*, and I should tell you that you won some kind of essay contest."

Then it sank in. My behavioral biochemistry essay had won! In the excitement of the moment, I have no memory of what I said, but according to Peter, my reaction was understated.

"You're kidding me! Sun of a gun."

Carol walked into the lab a few minutes later. Mad with excitement, I composed myself to ask matter-or-factly, "Guess who won the *Perspectives* essay contest?"

She, having clearly discounted the possibility that it could have been me, said in earnestness and without hesitation, "Who?"

"Well, who do you think?"

When the realization dawned on her, she became more excited than I appeared to be and proceeded to broadcast the news up and down the halls of the Shriver Center. Congratulations were immediately forthcoming and warm, especially from Marjorie Lees, which meant a lot to me.

At NRP the following Monday, I was faced with a ticklish political problem. Fortunately, I had just the previous week placed a copy of the revised manuscript, the one actually sent off in December, on Fred Worden's desk with the notation that I was submitting it as an entry in an essay contest, "not necessarily for publication." When Key Dismukes had published a couple of notes in the News and Views section of *Nature* months earlier, Schmitt and Worden had become very agitated that he had not cleared the material with them beforehand. Worden was on vacation in Arizona, so I waited a few days after he returned to give him time to find my note before springing the news on him that I had won the contest and that the essay would in fact be published after all. In the meantime, I went to Schmitt to thank him for his comments on the earlier draft and tell him that I had submitted the finished version, incorporating his suggestions as an entry in an essay contest. He said he figured that I had in a tone of voice that suggested I probably had wasted my time.

When Worden did find out I had won, he was generous and genuine in congratulating me. Nothing I had accomplished in my career to that point had made me feel better. Not least of the reasons, in my mind, was vindication of the worth of the Staff Scientists as scientists in their own right. I had now joined Key Dismukes and Parvati Dev in publishing work from my own thoughts and creativity while on the staff at NRP.[4] Therefore, I would be recognized, I was sure, for my award-winning accomplishment at the Stated Meeting of the Associates, the world's most distinguished collection of neuroscientists, to be held just two weeks hence. On the first night of the meeting, at the first gala dinner where introductions were made and announcements of notable accomplishments dispensed, I was beside myself with excitement. When the evening had ended, however, no one had mentioned my name. I had become the first Staff Scientist in the history of NRP to win a major writing award, and no one had thought it worth noting.

[1]   Irwin, C. C. and L. N. Irwin. 1979. A simple rapid method for ganglioside isolation from small amounts of tissue. *Anal Biochem* 94: 335–9.
[2]   Irwin, L. N., R. A. Barraco, and D. M. Terrian. 1978. Protein and glycoprotein metabolism in brains of operantly conditioned pigeons. *Neuroscience* 3: 457–63; Irwin, L. N. and D. M. Terrian. 1978. Glucosamine incorporation into rat cerebrum: effect of adrenalectomy, corticosterone, exercise, and training. *Pharmacol Biochem Behav* 9: 333–7.
[3]   Irwin, L. N. and C. C. Irwin. 1982. Development changes and regional variation in the ganglioside composition of the rat hippocampus. *Brain Res* 256: 481–5.
[4]   Irwin, L. N. 1978. Fulfillment and frustration: the confessions of a behavioral biochemist. *Perspect Biol Med* 21: 476–91.

34

The Decadent Pleasure of
a Gentlemen's Club

The remainder of the 1978 spring Stated Meeting was a mixture of the good, the boring, and the obscure. Falling in the latter category was Ted Bullock's farewell talk on the occasion of his "elevation" to emeritus status at NRP. Though he had vowed to me at the last Stated Meeting that he intended to be in the Amazon in order to skip his own graduation, he had shown up after all, out of respect for Schmitt perhaps or simply because, like the rest of us, he didn't want to pass away unnoticed.

The parallels between his last hurrah and that of Ross Adey were uncomfortably close. He, too, gave a fairly brief and elliptical speech, with obscure inferences that no one seemed to understand. And again, neither Worden nor Schmitt prodded him for elaboration. The second Associate in six months, and the second one to have played a huge role in the development of neuroscience, was passing from the scene at NRP with nothing but meaningless homilies spoken on his behalf. Fortunately for neuroscience, both Adey and Bullock would remain active for years after their NRP days had ended.

Sheets or Globs

In May, a meeting on recent theoretical approaches to neurobiology gave me a chance to meet the legendary Francis Crick, codiscoverer with James Watson of the double helical structure of DNA. With uncharacteristic boldness on my part, I introduced myself to him during the cocktail hour of the first evening and, after small talk to break the ice, asked him if it were true that he had recently observed that "the brain appears to be organized into cell collections that come in either 'sheets' or 'globs.'"

He laughed and confessed to a remark somewhat resembling my paraphrase. He expounded by explaining that he was still at the stage of asking a lot of naive questions. Lawrence Bragg, Nobel laureate and head of the Cavendish lab where Crick and Watson had worked, used to ask simple questions that everyone else was afraid to ask. That seemed to be his function at the Cavendish. As the meeting progressed, that appeared to be the role that Crick had assumed at NRP. Except that his questions were never truly naive; and his remarks, frequently punctuated by a cross between a hearty laugh and a sudden snort, were invariably incisive and insightful. Equal in intellect to Gerry Edelman and every bit as bold as Jack Cowan, both of whom were also in attendance, his presence dominated the meeting. But Mountcastle was back again as well, speaking softly and seldom, but when he did, cutting to the heart of the matter. With unaccustomed fervor, he implored the participants to try to come to grips with the fundamental crossroads at which neuroscience had now arrived—the juncture between single-cell activity and the behavior of neuronal populations. Would any amount of cellular detail ever tell us how cells behave in groups? Ross Adey and E. Roy John, who could have commented from their disparate points of view on this issue, were not there to do so.

The repartee between Edelman, Crick, and Cowan throughout the meeting was a joy to watch. Edelman and Crick, both with Nobel Prizes to their credit, enjoyed the license to say whatever was on their minds. Cowan, without a Nobel Prize, took the same license anyway. Leavened by prodding, almost brooding, questions from Mountcastle, the meeting at times reached a height of scintillation that I witnessed at no other time at NRP. And the fact that collective intellects as facile as those of Edelman, Crick, and Cowan could not answer Mountcastle's query was testimony to the intractability of the problems of higher brain function.

On this point, I was most interested in Crick's opinion. Is there something about the problem of brain function that makes it inherently more difficult to solve than the problem of gene function? Probably so, he replied, because the brain, unlike the gene, appears *not* to be resolvable into smaller parts that behave in isolation the way they do in the intact, integrated organism. There was much discussion during those two days about whether it is better to start with the intact system (behavior) and work down or begin at the lowest level of resolution (molecules) and build up. David Hubel's remark left the debate at a draw. "One does what works," he said. Pick a system that can give you answers to specific questions, and see what it tells you. The day for putting it all together was not yet upon us.

Dinner Like No Other

The evening between the first and second days of this meeting provided me, quite by chance, with one of the most extraordinary experiences I would ever have in Boston. It came about because Frank Schmitt and Fred Worden were thrown into a dilemma by the arrival in town of a Venezuelan financier, allegedly one of the wealthiest men in the world and a generous past contributor to the Neuroscience Research Foundation—the vehicle for private donations that financed the first-class airfares for Schmitt and Worden and the lavish entertainment expenses for the Associates that federal grants were prohibited from funding. While Schmitt and Worden were anxious to wine and dine the financier, they had the problem of what to do with a house full of eminent scientists that otherwise wouldn't take kindly to dinner at the local Howard Johnson's in Brookline. Their solution was to dispatch the party in its entirety to St. Botolph's Club, an exclusive Back Bay establishment for members only. I was recruited to be a chauffeur for Gerry Edelman and Vernon Mountcastle. By that accident of history only, I got into St. Botolph's Club for the first, and most likely last, time in my life.

Sensing that this could turn into something special, I deliberately sought a seat at the most interesting table I could find. Regrettably, Edelman and Crick were not seated together, and seats around Crick filled quickly; so I settled for the last available chair at the table with Edelman, Mountcastle, Jack Cowan, Werner Reichardt, and Barry Smith. As the wine began to flow and the exquisite dinner of broiled salmon was served, the five of us settled into convivial conversation on a plane to which I was not accustomed. Barry and I were passive listeners for the most part, as was Reichardt. But Edelman, Mountcastle, and Cowan kept up a spirited dialogue that entertained and intrigued us all evening. They talked of the famous people who had been their friends, like Lars Onsager and Warren McCulloch for Edelman and John Dos Pasos for Mountcastle. They argued politics, debated who was the greatest living mathematician, prognosticated on the fate of the feminist movement, told ribald jokes, and related touching stories. What struck me was not so much the level of conversation, though it was high by any measure, but the sense of fraternity that existed among five table companions who, ranging from a Nobel laureate to a nobody during the earlier hours of the day, were at this hour simply enjoying one another's company as men with views to air and stories to tell.

Taste of Cigars and Talk of Women

At no point in the evening was the feeling of camaraderie more intense than when we got onto a discussion of women. Past dessert and well into cordials, Jack Cowan offered me a premium cigar that I couldn't pass up. With all but Barry enjoying that most decadent of gentlemanly pleasures—a good smoke—we launched into an examination of our true feelings about the state of our female colleagues. There are two things in neuroscience that are certain, said Edelman. The first is that neuroscientists turn to philosophy as they approach old age. The second is that there are hardwired differences between the brains of males and females. To confirm the latter, he asked for a show of hands by those who agreed. All five of us raised our hands. (The scientific reality of this is now beyond dispute, but what it means is still controversial.) Edelman recounted a cocktail conversation with a woman whom he informed, following a lecture from her on the necessity of equality between men and women, that there are hardwired differences between the way brains of males and females are put together.

"She was dumbfounded but later had the grace to come back and tell me that she 'just realized that science extrapolates from the particular to the general.'"

There was talk about the lamentable difficulties that a woman faces in negotiating the multiple hurdles of getting through graduate school, raising a family, and dealing with societal expectations not conducive to scientific careers for women. There was the suggestion that women may be trying to achieve too much. Someone noted that women cannot have it all. In one of my few intrusions into the conversation, I said I thought that what women wanted to have was not so much all of it as simply a choice of it. Edelman predicted that the feminist movement would die out in about five years, when women realized they had lost more than they had gained. It was not the only observation at the table that evening that would turn out to be wrong in both its substance and its assumptions. But no matter. What counted for the moment was that we were a brotherhood, unencumbered by pressure for political correctness, free to speak of the way things are as if we really knew. I understood that evening the great appeal of private clubs, especially of clubs for men only; and for just a few hours, I loved it without apology, knowing that my populist roots would never let me feel truly comfortable in that atmosphere for long.

Poignant End to the Evening

On the ride back to the hotel, my companions settled into a more somber mood. Driving into town, Mountcastle had sat in the front seat beside me. We had talked about my career and his, about his book with Edelman and other matters, as if we were the equals that we weren't. I was touched by his evident concern for me as a person, rather than as a part of the wallpaper at the Brandegee Estate. (Indeed, the Associates as a group were always much more interested in and solicitous of the Staff Scientists than were our own in-house supervisors.) On the ride back to the hotel, however, Edelman and Mountcastle sat together in the back, sharing a few minutes of very personal conversation far different from the elevated rhetoric of the hours previously. They talked of growing older and of changing times. "My friends are dropping like flies," lamented Edelman—some dying, some divorcing, some dropping out of science. There was mention of the middle-age crisis. Mountcastle poignantly described passing through his several years ago, coincident with the death of his son. He had lost "a couple of years of research" but had eventually passed through it and was doing well enough now. Edelman said something about no tragedy befalling him "comparable to the loss of a son, but . . ." The remainder was drowned out by the noise of the Boylston Avenue traffic. Soon we were back at the Chestnut Hill Hotel. Thus the day, which had produced the most elevated scientific discourse and the most uniquely fraternal camaraderie of my years in Boston, ended on a quiet, poignant, and rather sad note.

35

Sweet Rush of Joy

Back in the real world, Carol and I were making slow but steady progress in the lab. She decided to go to the meeting of the Federated American Societies of Experimental Biology in Atlanta in June. This was not a meeting I regularly attended, but most of her friends and colleagues in biochemistry did, so it made sense for her to go while I stayed home with Anthony and continued to work. Also, she had done all the work on scaling down the ganglioside isolation procedure, so she was the natural one to present that work at the meeting.

I regretted that there wasn't more substance to this first paper that we would present as coauthors. We did have some good preliminary data on small samples of retina, tectum, and hippocampus to illustrate the utility of the methodology, but had not yet developed a complete developmental sequence for any of the tissues. It was adequate for a poster presentation at a meeting, I thought, but not yet worth a publication.

To my surprise, Carol returned from Atlanta with news that our poster was a big hit, had attracted a great deal of attention, and was obviously very timely. Many new people were beginning to work on gangliosides, and our simple, rapid, and economical procedure was something that many of them had been looking for. Richard Himes, her former mentor at the University of Kansas, told her that we should try to publish it. I was still dubious since, after all, it was merely a modification—and a fairly small one at that—of a method that really had been developed by Bob McCluer. But he, too, felt it was worth trying to publish, so with that encouragement, we wrote it up for *Analytical Biochemistry*, a short article of only about three pages. Even after it went to press, I was not particularly impressed by it, but it would give us that much-needed first publication fairly soon after our arrival at the Shriver Center. It was accepted, subject to a few minor changes. From the beginning, it was obvious that it would be a big success because we received a ton of reprint requests. In time, it became the most widely read and often-cited work that Carol and I ever wrote.[1]

Preposterous Plan to Discover the Earliest Ganglioside

In every developmental tissue that we looked at, we were now seeing the same pattern of transition from smaller, simpler to larger, more complex gangliosides as the tissue matured; and the appearance of the more complex gangliosides correlated nicely with the end of cell multiplication and the completion of cellular differentiation in the brain. Once embryonic cells decided to become neurons and decided what type of neurons they would become, they started converting the simpler gangliosides into more complex ones. The decision was presumably made at the genetic level, in response to some signal relating to the position of the cell in the overall nervous system, the cell that it was next to, or some other factor—the consequence being a new set of enzymes that would promote the addition of more sialic acids to the core of the molecule. Our developmental work on gangliosides, in other words, provided a good model for studying the way the nervous system matures at the metabolic level. Understanding that, in turn, was a prerequisite for figuring out what goes wrong when abnormal neural development results in mental retardation. It felt good to be working on something of such obvious potential human benefit.

One question that remained unresolved was the nature of the ganglioside pattern in the *very earliest* embryonic nervous tissue in mammals. This had been technically difficult to answer because those earliest stages occur during the first few days of gestation in utero when the embryonic brain is tiny. From the pioneering work of Kunihiko Suzuki, it appeared that the earliest major ganglioside in the rat brain was one named GMl—a ganglioside with a four sugar backbone and one attached sialic acid (Fig. 3). Yet our work, and that of others based on samples earlier in development, was indicating that in chick embryos, the earliest major ganglioside was GD3—a ganglioside with a backbone of only two sugars, but with two sialic acids attached. Thus, the true parental ganglioside pattern in the earliest mammalian embryos remained an open question.

Just as we were thinking through this issue, I got a call from Dan Michael, a former undergraduate technician for Robin and me who had since graduated and proceeded to medical school at Wayne State. He and his fiancée, Melinda Raab, wanted to spend a little time in Boston. Therefore, he wondered if I would be willing to have him come work with me at the Shriver Center on a project that he could get done in about three weeks. I told him I had never accomplished anything of substance in science in only three weeks, but if anyone could, it would probably be him. And of course, Carol and I would be happy to have him as a visiting scientist in our lab. Trying to pin down the early ganglioside pattern in a mammalian embryo in such a short period of time would be a long shot, but it was a specific and important question with a limited objective. We set his arrival date for early September, and I ordered the rats to give us an early developmental series of fetuses to coincide with the time he would be there. Dan would provide free skilled labor, and Carol would be available to supervise, so I figured the effort would be worth it.

Frivolity and an Ill-Fated Offer

Meanwhile, at NRP, changes were in the works. Having Barry Smith as Program Director had been a lifesaver for the other Staff Scientists and me because he was very sympathetic to our discontents.

But he had been unable to implement the reforms that the organization needed. By early spring, Key and I were predicting that he would soon be gone, and by late spring, he had agreed to accept an excellent position as director of a neurosurgery lab at NIH.

June and July at NRP were consumed with enthusiasm for the cerebral cortex. Schmitt assembled us all to prepare him for a visit to Max Cowan at Washington University in St. Louis. We spent a lot of time discussing the cortical columns that Mountcastle had proposed as the functional units of the cortex and which had become an important basis for Edelman's theory of multiple reentrant information processing. Schmitt seemed obsessed with defining the precise physical dimensions of the columns, and I kept arguing with equal obsession that the precise dimensions were irrelevant. We finally assembled a list of sixteen questions for Schmitt to take to Cowan—all of them elementary, in my view, and answerable by any Staff Scientist entrusted with the project in a couple of hours at the library. When Schmitt returned from St. Louis with a learned discourse from Cowan on all the questions put forth, Worden congratulated us for the impressive homework we had done in-house. If we had really been allowed to do the work ourselves, we would have saved an airfare to St. Louis and six hours of Max Cowan's time.

But Schmitt was in a good mood those days. One morning, I was wearing a shirt with a brightly colored floral pattern, sitting with my back to the door. He walked up behind me and swatted me on the shoulder with a manila folder, then recoiled in feigned embarrassment, exclaiming, "Oh, I'm sorry! I thought I saw a bee there!" At a later point, he was launching into another one of his grand misplaced analogies between neuroscience and molecular biology, when he interrupted himself suddenly with, "Well, never mind. Lou doesn't like for me to do that." I laughed with everybody else, complimented by this unexpected evidence that I really had an effect on his thinking.

Schmitt was not a humorless person, and he certainly wasn't devoid of acumen in both scientific and personal matters. He was a talented writer and a gifted integrative thinker. Somehow either his upbringing or his experience in the academic science of Germany in the late 1920s had left him obsessively reliant on reference to authority, when in reality, he could have commanded more of it on his own.

With something that in retrospect I can only liken to a death wish, I decided to ask Fred Worden if I could be made interim Program Director until a permanent replacement for Barry could be found. I made this request on August 3, my mother's birthday, perhaps in unconscious obeisance to her opinion that I was never assertive enough. I guess at that time I was still struggling to find a way to matter at NRP, not to mention that my employment prospects otherwise were dismal. Having managed the files on all the impending topics and work sessions—a task originally assigned me by Worden—I was sure I could handle the mechanics of the job. My guess is that Worden was dumbfounded at the suggestion, but to his credit, he composed himself to simply say that he had thought we would try to get along without a Program Director for a while, but he would talk it over with Schmitt and get back to me.

Twelve days later, Worden called me into his office and told me in a very nice and respectful way that he had decided "not to take [me] up on [my] offer." He went on to give a couple of rational-sounding reasons that were too facile for credibility. In a way, I was relieved. This convinced me that I had to leave NRP by the end of the year, with or without another job. I would devote myself exclusively to my own interests from that point forward and to my one remaining NRP project: salvation of the retinotectal specificity work session.

Back to the Connection from the Eye to the Brain

As edited manuscripts came back over the spring and early summer from participants in the retinotectal neurospecificity work session, I formed three conclusions: (1) the vast majority of participants were fair and open-minded scientists interested in seeing their research field rise above the petty acrimony that had characterized its origins; (2) Marcus Jacobson was an egotistical prima donna without whom the final publication would be better off; and (3) Sansar Sharma was a key to understanding what Mac Edds had really intended the meeting and the *Bulletin* that would come from it to be.

With mostly gentle and some persistent persuasion, I convinced everyone except Jacobson to return to the project that most of them had given up for dead. The more I got to know them through correspondence and phone conversations, the more I came to like them and respect their long, tedious, and technically difficult experiments, conducted as they had to be in such a highly charged field. Especially the younger ones— Myong G. Yoon, Ray Lund, Susan Udin, Ronald Meyer, and Tim Horder—earned my respect. I had been less familiar with the work of Michael Gaze before starting the project but came to admire his work too as I learned more about it, even if Gaze, I sensed, never felt full confidence in me. Of Jacobson, I had the same regret that we all suffer when learning that a person whose work we had greatly admired in the abstract turns out to be so unpleasant as a human being. Including him in the *Bulletin* would almost surely have generated such rancor that the project's completion would have been in doubt. But it was he, not me, who made the decision that he would be excluded.

Poring through the background material that had led up to the November 1975 work session, I learned how close Sharma and Edds had worked together in planning it and therefore resolved to go to New York to talk to Sharma in person. As friendly and cordial in person as he had been over the phone, he filled me in on a myriad of helpful details and illuminating anecdotes about various incidents and people. None was more fascinating than his description of how he had come to discover the compression of the retinal projection onto an experimentally reduced area of tectum.

In the early '60s, based largely on the work of Roger Sperry, it was generally assumed that retinal cells make precise connections with one, and only one, target cell in the tectum. Thus, if the optic nerve were crushed and half of the retina were removed, when neurons from the surviving half of the retina regenerated, they would form connections with target neurons over only half of the tectum. By the same token, if half the tectum were removed, then only half of the retinal field should remain connected. Sperry's experiments had shown this indeed to be the case.

Sharma and Gaze decided to repeat this experiment in goldfish. After the surgery was completed in March of 1964, Sharma took off for India by automobile, a trip that turned into a six-month odyssey. By the time he got back to the lab in Edinburgh, it was November, much longer than required for regeneration of the optic nerve in his operated goldfish. When he subsequently mapped the retinotectal projection in these animals, he was astounded to discover that the entire retinal field had formed connections *compressed* into the remaining half of the tectum. The connections were all lined up in the right order, but were simply twice as close together as they would be in the normal brain. In other words, given enough time, retinal connections to the tectum would rearrange themselves, in proper topological sequence, to fit the available space. This showed for the first time that neuronal fields form connections *as distributed systems*, not according to precise one-on-one instructions.

Roger Sperry reviewed the paper submitted by Sharma and Gaze on retinotectal compression and offered the opinion that both senior and junior authors should read a textbook on elementary embryology before trying to publish such erroneous conclusions. Eventually, the paper got published anyway, and Sperry, to his credit, told Sharma, "You were right, I was wrong."

"No, Roger," Sharma replied, "we were both right." Subsequent experiments confirmed that indeed both results could be obtained, depending on how soon after regeneration the projection of the retinal field was tested. It was a clear indication that nerve cells don't find their connections by looking up the precise address of their targets, at least over the long run. Rather, they simply find their proper place in the neighborhood in relation to their neighbors, however near or far away from them they have to settle. Given this complexity and the cleverness of an ever-growing army of graduate students, the experiments over time had become more complicated and confusing. It was out of an effort to clarify this muddled situation that the work session at NRP had been conceived. I returned to Boston, convinced more than ever that publication of the proceedings of that work session would be a service to neuroscience. There was still the matter of dealing with Gaze and Schneider, but I resolved to do it.

Gaze had sent Schneider a draft write-up of that portion of the *Bulletin* for which Gaze was responsible in November of 1976, a full year after the meeting had been held. Schneider did not respond. Yvonne Homsy (publication coordinator at NRP), George Adelman, and Fred Worden had taken turns in prodding Schneider to write up his portion of the meeting. Each of their phone calls was usually met with a promise of a manuscript within a month or so, but no manuscript was forthcoming. On Adelman's recommendation, Worden had agreed to assign me as a coauthor on the *Bulletin*. Though it cut against the grain of NRP's philosophy that Staff Scientists should be seen but not heard, Worden recognized that it was the only chance this particular work session had of ever seeing the light of day. Had Mac Edds not died, casting the meeting essentially as a memorial to him, I am quite certain that Worden would have killed the project. Instead, he backed me forcefully in this instance, giving Schneider an ultimatum to go forward with me as collaborator or be left behind. At every step, Schneider gave ground grudgingly, but talked just enough to keep us from cutting him out entirely. Finally, in July of 1978, after a lengthy meeting with me, he produced a manuscript much reduced in scope from his original responsibilities, but of good quality on his section.

Big Bugs and Novel Gangliosides

At the end of the summer, Marion and Chester Jacewicz—long-time friends from Carol's school days in Tulsa—came to see us; and we took off for a fabulous family vacation together to Bar Harbor, Maine. It was a marvelous trip, complete with a moonlit night beside the ocean, exquisite seafood dinners overlooking rocky shores, and leisure progress down the backroads of Maine at the forty-miles-per-hour maximum speed that our now frail and aged VW camper could muster. So fond were our memories of that trip that the next time they came to visit, I brought lobsters home for a memorial dinner. By then, their youngest son, Stefan, was just starting to talk. Born and bred in Oklahoma where insects are common but shellfish are not, Stefan took his first look at a lobster in our kitchen sink and stammered, "Big bug!"

The day after Marion and Chester left, Dan and Melinda arrived. Just why Dan and I were really going to attempt a publishable experiment in only three weeks still escapes me. Surely, it was only because Dan had suggested it, and Dan was the type of person whose folly could not easily be dismissed. He had completed the requirements for a major in psychology at Wayne State in only three years. During his junior year, he heard of Robin and me through his uncle, Don, and started attending our evening neuroscience seminars. Impressed by our tolerance for the offbeat and philosophical side of science, he asked to work for us the following summer. We rewarded his interest by making him a laboratory technician devoted to the task of running thousands of tedious protein assays. (We were analyzing a lot of fractions in those days, so our students had to earn the right to be philosophical the old-fashioned way.) Immediately upon his graduation in 1975, he was admitted to medical school, so he, at this point in time, was starting the last year of his medical training and had chosen to do a research rotation, if you could call it that, with me.

Carol and I outlined the project that we had in mind. We would collect embryos at various stages of development from pregnant rats, pool the miniscule amounts of brain tissue we could obtain from them at the earliest possible age, show him how to isolate gangliosides from the tissue and run the chromatograms, and see what happened. Given the short time available, it would be a one-shot attempt with no margin for error. I would show him how to collect the embryos and dissect out the brain tissue. Carol would be there to help him with the biochemical procedures. Otherwise, he would be on his own.

Dan threw himself into the work of the lab. A quick learner, he was soon doing all the biochemistry himself, under Carol's watchful eye. We got the embryos at appropriate stages and appeared to dissect out enough brain tissue for the analysis. It just didn't look like there was going to be enough time. Carol insisted that he run some practice thin-layer chromatograms first. They did not turn out well. We had to get sharp enough resolution to reveal whether that earliest ganglioside band from the embryonic tissue was going to travel right next to the GM1 reference standard or travel below it, where GD3 ought to be. It came down to the final day. All the gangliosides from the earliest embryos we could get would have to go on one plate, and it would have to be run on the Friday before the Saturday they had booked their return flights to Detroit.

For some reason, that Friday, I had to be away from the Shriver Center, so I didn't learn the outcome till I got home. When I walked in, Dan and Carol were glum.

"How did it go?" I asked.

"Well, I don't know." Carol sighed. "You can see for yourself. The plate is over there." She pointed to the small square of glass coated with silica gel, with one lane for standards and several lanes for the different embryonic stages.

I picked the plate up, fearful of what I would see, and turned it to the front. There, lightly but clearly in the lane for the earliest developmental stage, was a single purple band distinctly below the GM1 standard. And there were no bands in the slowly migrating region, where the complex gangliosides would have been. The first ganglioside to appear in mammalian development was GD3! In that instant, I realized that the three of us knew a fundamental fact of nature that no one else in the world knew. We had made an original discovery. I sat down slowly, then looked up at the now broadly grinning Dan and Carol and was flooded with a feeling of joy.

Science calls to its practitioners for a number of reasons. We believe that the universe is ordered by laws that can be studied by objective methods, and we prefer an ordered world. Many of us enjoy the manual effort of carrying out the experiments themselves, working with our hands as well as our heads. Most of us derive pleasure from "playing mind games with nature," as Dianna Redburn had once put it. But all of us, without exception, are driven by the hope that one or two times in our lifetime, or a few if we are exceedingly lucky, we will experience the sweet rush of the joy of an original discovery. In the life of the intellect, no other feeling comes close.

Leaving with Mixed Emotions

Dan and Melinda returned to Detroit. Carol and I carried out a few more experiments to prove that the single band we had seen was in fact GD3.[2] And I prepared to leave NRP for a future unknown. Nothing could have been a stronger indication that it was time to do so than the fact that at the fall Stated Meeting of Associates, Fred Worden had introduced the entire science staff, including the staff writers, except for me, whom he forgot until prodded from the head table (I think by Schmitt himself).

On December 14, 1978, it was my turn to go through the ritual of the goodbye party. I was called to the library on a pretense by Schmitt, where all but one of my favorite secretaries, Bonnie, had assembled with sherry, Lancers Rosé, potato chips, and dip. Key Dismukes presented me with the institutional gift of several classical recordings, including Pachelbel's "Canon," which I listen to with pleasure to this day. Characteristically unable to wait for Bonnie to finish typing a letter, Worden launched into a two- or three-sentence toast about how great it had been to have me at NRP, etc. He did mention that I was the only Staff Scientist to have won a writing competition while in residence, and I appreciated his remembering that, though I would have appreciated more his remembering to say it at a Stated Meeting. Schmitt also had a few kind words to say, then offered to write me a letter of recommendation for a faculty position in biophysics in California. He then asked, parenthetically, if I considered myself a biophysicist or molecular biologist or something else, giving every indication that he still knew little about me.

Notwithstanding my nearly constant feelings of dissatisfaction from the day I had arrived at NRP, I could not leave without emotion. Whatever I thought of the institution, I regarded everyone there my friend. Certainly all the Staff Scientists—Bruce Brandt, Parvati Dev,

Steve Dennis, and Key Dismukes, especially. Key was probably the best writer and one of the best thinkers to ever pass through there. Yvonne Homsy had been my chief collaborator as a technical writer. George Adelman was an unstinting supporter and the closest lasting friend that I would take from my days at NRP. Even Fred Worden and Frank Schmitt, by far the chief targets of my frustration and criticism, had been unfailingly considerate and kind to me at a personal level.

All this, I had intended to say. But as in the case of my departure from Wayne State, I was not invited to give a going-away speech. I appreciated the party, but it was not a big deal. When Barry Smith had left back in the summer, we had a dinner party at the Worden's estate in Weston. When Parvati Dev left, Schmitt was effusive in his praise. Even Bruce Brandt at least got pizza at his going-away party. For me, we had a few chips and dip, and went home.

[1]   Irwin, C. C. and L. N. Irwin. 1979. A simple rapid method for ganglioside isolation from small amounts of tissue. *Anal Biochem* 94: 335–9.

[2]   Irwin, L. N., D. B. Michael, and C. C. Irwin. 1980. Ganglioside patterns of fetal rat and mouse brain. *J Neurochem* 34: 1527–30. We submitted the paper in January 1979, but it got lost in the editorial process and didn't appear in print till mid-1980. Unknown to us at the time of our discovery, Ephraim and Ziva Yavin, at the Weizman Institute in Israel, were working on the same problem. Their paper[3] was published in early 1979, about the time that our paper was first submitted to the *Journal of Neurochemistry*. So, in fact, the Yavins discovered the same thing we did at the same time, if not slightly before us. But we didn't know that at the time, so it posed no check on our elation.

[3]   Yavin, E. and Z. Yavin. 1979. Ganglioside profiles during neural tissue development: Acquisition in the prenatal rat brain and cerebral cell cultures. *Dev Neurosci* 2: 25–37.

36

Getting a "Qualified Success Plus"

I continued to work on the retinotectal neurospecificity issue of the *NRP Bulletin* after my official departure in December 1978. Gaze made it difficult, but in the end, he came around and was helpful. Kevin Hunt was argumentative and dilatory but eventually made a useful contribution. Gerry Schneider, on the other hand, drove me to distraction, and it isn't far from the truth to say that the *Bulletin* finally appeared in spite of rather than because of him. At long last, volume 17, number 2 of the *Neurosciences Research Program Bulletin* was published in April of 1979. Entitled "Specificity and Plasticity of Retinotectal Connections" by Mac V. Edds Jr., R. Michael Gaze, Gerald E. Schneider, and Louis Neal Irwin. Its appearance marked the second resurrection that I had engineered at NRP and was probably one of the most difficult but also most satisfying publications of my career.

Much of the first part of the year, I spent writing a grant proposal to NIH on hippocampal gangliosides in response to a call for proposals related to local neuronal circuits. This was clearly evidence of the influential hand of NRP, which had raised the topic to a high level of visibility. I proposed to study the expression and turnover of gangliosides within specific regions of the hippocampus that could be identified as to cell type and connectivity, I would use developmental changes, drug and surgically induced ablations, and hormonal stimulation as different ways of manipulating the morphological and physiological state of a discreet brain region. It was a grab-bag of approaches, but the hippocampus was hot, and I thought I could make a persuasive argument that the experiments collectively would shed light on the structure and function of some well-known local neuronal circuits at the molecular level.

Into the Void

When I left NRP in December, I was left with no salary at all. Our entire family income derived from Carol's salary at a postdoctoral level on my NSF grant, which was due to expire in July. Ironically, my research in partnership with Carol was going better than it ever had, and we were beginning to make a small name for ourselves in neurochemistry. That I should be casting about for a way to pay the mortgage and salvage my career at this stage became a growing source of depression for me.

The nadir came just as I was leaving for the American Society for Neurochemistry meeting in Charleston, South Carolina. I was informed that my grant application of the previous year to NSF, a proposal to study phylogenetic variation in ganglioside patterns as a clue to their biological function, had been turned down. Too much of a "fishing expedition," said one reviewer, who clearly had no understanding of evolutionary theory or the importance of comparative biology. I swear to God, if Darwin had submitted a grant proposal to fund his voyage on the *Beagle*, it would have been turned down on grounds of being just another fishing expedition.

To add insult to injury, the rapid transit line that normally could be relied upon to get me from Newton Highlands to Logan Airport within two hours was exceptionally unreliable on the day of my departure, and I missed my flight, causing me to sit at the airport an extra four hours and brood over the continuing downward spiral of my career. The meeting did little to lift my spirits. The poster that Bob McCluer, Eric Bremer, Carol, and I coauthored on gas-chromatographic separation of gangliosides from different segments of the retinal field was given a generally underwhelming reception. On two straight nights, I looked for friends to have cocktails or dinner with, and failed. The first night I had a scotch and soda alone in my room, then went out for dinner alone. The second night, I had a couple of scotches and sodas in my room and never made it out to dinner at all. On the final evening, I did run into some acquaintances in the hotel bar, to which I had been drawn by a band belting out "Luckenbach, Texas." Among the group was a young woman whose name I never learned from a research lab at NIH. By her silence and a few things she said, I gathered that she felt worse about herself than I did about myself just then, so my friends and I persisted in drawing out more information about her background, research, goals, and so forth. By the end of the evening, I think we had made her feel more important, and that was the best—really, the only—good feeling I had all week in Charleston. Paradoxically, meetings can sometimes be lonely, and this was one of the worst for me.

Things Change for the Better

But it was just at that time that fortune began to turn. Upon my arrival back in Boston, Carol greeted me with the news than Sonya Sobrian, the program director for my NSF grant in Washington had called to say that the agency was going to extend the grant for six months, with a 10 percent increase, to give us time to submit a new or different application. This was wonderful news because it ensured that we would have some income and could keep our lab open, at least till the end of the year. As it turned out, this unilateral act of generosity on the part of a bureaucrat[1] in Washington bought us just enough time to save our careers.

In May, I presented a seminar at the Shriver Center entitled "What's Happening in the Retinotectal Field?" I described recent work in which retinal tissues incubated in vitro had been found to shed proteins into their extracellular environments. I speculated that these molecules, rather than representing breakdown products or the residue of broken cells, might be actively secreted as a biological signal of some sort. And if so, they might have receptors, which could, of course, be gangliosides. Carol had the idea of immobilizing gangliosides in a column, incubating retinas in vitro, then passing their extracellular fluid through the column to see if any of the secreted molecules would stick to the gangliosides in the column. Gary Schwarting, a colleague at the Shriver Center, had used such a technique to isolate ganglioside-binding antibodies, so he showed us how to set it up.

By June, I was fitfully preparing for an assignment associated with the second Gordon Conference on macromolecules and behavior. Two years earlier, I had suggested to Ed Bennett, the program committee chairman for this meeting, that instead of a keynote speaker following the traditional sumptuous lobster dinner on Thursday evening, we should have a panel discussion on the basic conceptual issues that plague the field. He thought it was a grand idea and, after reading my essay in *Perspectives in Biology and Medicine*, decided that I would make an ideal chairman. Also exercising his right as head of the program committee, however, he appointed thirteen members to the panel. I felt right away that thirteen was too many, but did not protest loudly enough. The general idea of the panel discussion must have been timely because twelve of the invitees responded to my preliminary correspondence with varying degrees of enthusiasm.

Twelve was still too many, and I worried excessively about keeping the discussion under control. My opening remarks on the evening the panel convened placed emphasis on keeping everyone to their stated time limits, with several humorous suggestions about penalties to those who didn't—like being sentenced to the hallway to listen to Yigal Ehrlich talk for half an hour. The laughs I got during my introduction loosened us all up, but once the serious discussion began, the tenor proceeded downhill. There were too many participants, too many questions, and too little focus. What incisive follow-up could have been promoted by me was not because I was concentrating on keeping control so everyone could have their say. It was a democratic discussion, but a flat one that never came to grips with the issues that needed illumination so badly. When finally the audience was invited to contribute, the pace and vitality of the commentary improved. David Wilson suggested that next time, participants be judged by *The Gong Show* standard: those panelists who offered either bland or obscure observations should be gonged into silence, in the fashion of the popular television show of the era. Lynda Uphouse, a fellow Texan whom I had just come to know and appreciate, pointed out that we talk as though we are studying the same thing (learning or behavior) when, in fact, we are studying many different things but calling them the same. Willem Gispen upbraided us for being too gentle with one another. More comments from the floor enlivened the proceedings, but it was too little too late. I was very disappointed.

In the rec room postmortem over beer and bourbon that followed, the reviews were mixed. Ed Bennett gave it a "qualified success plus," and others commended me for organizing a good program under trying circumstances. But some whose opinions I respected were candid in their disappointment. Bill Greenough, ever able to spot the cloud within the silver lining, along with Stephen Arch and Paul Gold, found the discussion stilted, limited, and inconsequential. In my heart, I agreed. At the time, I saw it as one more failure to come to grips with the central issues of behavioral neurochemistry. Now in hindsight, I wonder if there ever *were* any central issues. In a curious way, the failure of that seminar was vindication of the conclusion in my *Perspectives* essay that the meaningful questions, for the time being, were the smaller ones. Two years later, when we once again assembled at Brewster Academy in Wolfeboro, our Gordon Conference was renamed "Neural Plasticity," in frank admission that behavioral neurochemistry as we had known it, in response to Robin's prophetic question, had indeed died as the decade of the '70s was winding down.

---

[1]   Sonya Sobrian was much more than a mere bureaucrat. She was a faculty member at Howard University who carried out solid research in psychopharmacology for several decades. I have never, and will never, forget the small but critical role she played in saving my career.

37

First Night

In the twilight now of my years of devotion to the idea that the molecular mechanisms of behavior could somehow be unraveled, I was curiously energized by a series of unexpected successes. Carol's idea of passing retinal incubation medium through a column of immobilized gangliosides had indeed resulted in the retention of a tiny amount of material. We had labeled the retinal molecules with a radioactive tag, and the amount of radioactivity sticking to the column was barely twice above baseline, but it was a perceptible blip, and hadn't messenger RNA been first observed with a signal that small?[1] She ran the fractions that had stuck to the column through a gel gradient by electrophoresis to separate large molecules if they were present. Upon de-staining the gel, we saw three light bands, clustered around the 50,000 molecular weight region, meaning that at least three different sizeable proteins had bound to gangliosides. We were beside ourselves with excitement, seemingly on the verge of a second genuine discovery in little over a year. Unfortunately, in the tumult of the months and years that followed, we got sidetracked from this project. Others would eventually confirm the reality of the phenomenon and go on to characterize the ganglioside-binding proteins.[2] But this was future history that in no way dampened our enthusiasm at the time.

Career Calamity Averted at Last

Then on the morning of October 16, Marjorie Lees came beaming into my office at the Shriver Center, fresh back from the NIH Council upon which she sat in Washington periodically to review the latest recommendations for funding from the study sections. My grant proposal on hippocampal gangliosides had been given a decent priority score by the study section, though not enough to fund it under ordinary circumstances. But the program officer at NIH had picked it out as a special case because it addressed specifically the call for proposals related to local neuronal circuits and because it represented "a new career direction for the principal investigator." He recommended funding, and though Marjorie was ethically bound not to disclose the decision of the NIH Council, she made it clear that she was not telling me that funding for my grant had been approved. So the scientific team of Irwin and Irwin would survive after all, and maybe, if the fates were kind, even flourish over the longest grant period I had ever experienced: three whole years into the future.

This all came just as the executive committee at the Shriver Center had voted to promote me to associate biochemist and provide me with a secure salary for the foreseeable future. And a little trifle of a prize came to me on October 22, when I received in the mail a check for seven dollars in payment for my first short story, a tale of seven hundred words about a little boy and his lost kite that the magazine to which I had submitted it (oblivious of its Seventh Day Adventist sponsorship) had changed by a twist in the ending into a homily, in exchange for the one cent per word that they paid me.

As if the above were not enough, on the last day of October, I received a call from Simmons College, a small private college for women in the Back Bay of downtown Boston, inviting me to interview for a tenure-track position in the Department of Biology. The day of my interview, November 14, was cold and dreary; but my spirits were high, and the interview went well. I liked instantly everyone I met, but especially Sandy Williams, the chair, and Karen Loehr, a young assistant professor and head of the search committee.

Two days later, I was in the middle of tedious dissections of embryonic chick retinas at the Shriver Center when Sandy Williams called to offer me the position of associate professor of biology at Simmons. Suppressing my relief and happiness, I promised to get back to her after the weekend. Much as I wanted the security of a faculty position, the confidence that Bob McCluer, Marjorie Lees, and other colleagues at the Shriver Center had expressed in offering me a stable position there was compelling; and our research had just taken an exciting and significant turn. I hated to give any of it up and wanted time to figure out a way to have it all. Besides, there was another consideration, for the call from Simmons was not the only one of importance that we received that day.

More Good News

Our professional fortunes and failures in 1979 kept us on an emotional roller coaster throughout the year. Superimposed on these public events, however, weighed the very private and personal decision of whether to have another child. It had hardly seemed wise when our careers were in such flux, but with growing success at the Shriver Center, Carol especially had leaned toward repeating the rewarding experience that having and raising Anthony had turned out to be. I, on the other hand—feeling that my parental skills, like my panel discussion at the Gordon Conference, had been at best a "qualified success plus"—was not so sure. We decided to let nature decide, and nature didn't take long. A call from the Harvard Community Health Plan on that same fateful Friday informed us that Carol was indeed pregnant.

Returning to work on Monday, I informed Bob McCluer that I had been offered the position at Simmons. While emphasizing how grateful I was for his faith and confidence in me, I nonetheless confessed that I had grown so weary of living on the edge of professional uncertainty that I was inclined to accept the offer. To my relief and somewhat surprise, he congratulated me enthusiastically and urged me to take the job at Simmons, on condition that I stay involved to some degree with the Shriver Center. This was easy to agree to, as I had wanted to do so anyway. The grant would stay at the Shriver Center where, with Carol's help, we would continue to do research for three more years as a team.

But my primary affiliation would henceforth become Simmons College. I would end up teaching there for over a decade. Our second son, Brian, would be born in Beth Israel Hospital, a long block away from the office at Simmons where I slept fitfully a few hours the day he was born. Sandy Williams would become a valued colleague and close friend. Others, like Rachel Skvirsky, Joel Piperberg, Dick Nickerson, and Jane Lopilato, would likewise come to mean a lot to me. And for all the wonderful people I had known through the years, few would become a closer friend in time than Karen Loehr.

That all lay in the future, however. On the last day of December, I moved my office from the Shriver Center in Waltham to Simmons in the Back Bay. I did it alone, a labor of many hours that left me bone-weary. I wanted to go home and go straight to bed. But I had reached a juncture of great moment, and I couldn't bear to pass through it without a gesture of some significance. I had made a commitment to take a stand in a northern city in a climate I could barely abide because my family was growing, my science was changing, and the time had come to settle down. I would go to First Night, the uniquely Bostonian celebration of the onset of the new year, to pay fealty to what had finally, truly become my new home.

Finally Home

From a musical chorale at Trinity Chapel to an organ concert at the Arlington Street Church, I moved along finally with fifty thousand others to the edge of the harbor to watch the final gasps of the decade ignite in fireworks over the bay that Fred Samson's ancestor had entered some 360 years earlier. I thought of Samson, who had tolerated my deviations from the safer side of science with misgivings but support, and of Ungar, whom Samson despised but who had inspired in me those very deviations. That morning in Houston beside the refrigerator with the test tube that had fired my imagination seemed long ago;[3] the taste of fame had been so strong; the struggle for scientific respectability so long; the ultimate disappointments so deep.

But there had been moments. Like when I couldn't wait to get to work every morning at the Baylor College of Medicine in the summer of '66. Or the aroma of roasting coffee over the upper West Side of Manhattan. In Detroit, I had watched a couple of students stumble into graduate school, then ultimately emerge as first-rate scientists, in part because I had been there for a season in their lives. There was the morning of the call from *Perspectives in Biology and Medicine* and the evening in the fog-shrouded restaurant at the top of the Prudential Building in Boston to celebrate the call where, over wine with a tremulous voice, I had shared with Carol the unfathomable satisfaction that however little else I ever accomplished, this was a prize that no one could ever take from me. The evening I marched down the hill and into the stadium in Lawrence to receive my doctoral degree; the afternoon I first saw a new ganglioside in the embryonic brain of a rat that no one else knew was there; the morning that Carol and I first looked at the data showing evidence of proteins binding to gangliosides, discussed while still in bed because one of us had gotten home too late the night before to talk about it—each of these alone would have been worth the toil of years and a myriad of failures. To experience them all within a decade and a half made me think how fortunate I had been after all.

And what of the people who had enriched me so? Kay first of all, who had seen and believed in my promise; Penny, who had remained my friend when I had failed her; Bob and Muriel, who had saved my life in more ways than one; Marilyn and Jerry, who were always there at the critical points; Dorothy, who had given me the gift of music, and Robin, who had taught me to touch others; Dianna, who never seemed to doubt my worth; and Carol above all, who had made me more than I would ever have been without her.

No one had learned how the brain stores memories by that New Year's Eve of 1979—least of all me. The promise of discovery that had fueled my ambition from my earliest stirrings of scientific curiosity had, for the most part, not come to pass. But in the glare of the fireworks in the frigid night air, I was warmed by the thought that the search had been worth it. The light and noise died out with abruptness, and I headed for the Government Center station to catch the last train for Newton, aching with fatigue and anxious to get on with the rest of my life.

[1]   Judson, H. F. 1979. *The Eighth Day of Creation*. New York: Simon and Schuster, Touchstone Edition, p.440.

[2]   Watanabe, K., S. Hakomori, M.E. Powell, and M. Yokota. 1980. The amphipathic membrane proteins associated with gangliosides: the Paul-Bunnel antigen is one of the gangliophilic proteins. *Biochem Biophys Res Commun* 92: 638–646.; Schengrund, C-L., B. DasGupta, and N. J. Ringler. 1991. Binding of botulinum and tetanus neurotoxins to ganglioside GT1b and derivatives thereof. *J Neurochem* 57: 1024–1032; Sonnino, S., V. Chigorno, M. Valsecchi, M. Pitto, and G. Tettamanti. 1992. Specific ganglioside-cell protein interactions: a study performed with GM1 ganglioside derivative containing photoactivable azide and rat cerebellar granule cells in culture. *Neurochem Intl* 20: 315–321

[3]   Irwin, L. N. 2007. *Scotophobin: Darkness at the Dawn of the Search for Memory Molecules*. Lanham, MD: Hamilton Books., p. 4.

POSTLUDE

38

The Known and As Yet Unknown

The preceding narrative covered roughly the decades of the 1960s and 1970s because that was the period during which (1) fundamental questions about the brain mechanisms of learning and memory came into sharp focus and began to be illuminated, (2) the preeminent societies for neuroscience and neurochemistry with a worldwide impact were formed, and (3) I became a scientist and launched a career through the personal and professional failures and successes of early adulthood.

Neither neuroscience in general nor my personal or professional life came to a standstill in 1980. Remarkable advances have been made in the study of brain function in the years since then, and I have been privileged to see those advances and participate in them to a degree. But 1980 serves as a convenient baseline for what we have learned since then and what we still don't know. The following is a brief summary of the fate of the major issues that confronted us in those two momentous decades that defined modern neuroscience.

What We Know Now

Cell assemblies are a reality. In 1949, Donald Hebb proposed that information in the brain is manifested through defined neuronal circuits. He further proposed that the contiguity of groups of cell assemblies give rise to phase sequences, which are the cellular manifestation of cognitive functions that animals experience as imagery, perceptions, emotions, and (in humans and probably other neurally complex species) thoughts and foresight. By 1960, the reality of cell assemblies and phase sequences was a commonly held working assumption, since regional specialization and topological organization in the brain suggested that information is processed in the brain through defined circuits that correspond to distinctive information. Discovery of the specificity of retinotectal connections in the visual system, of spatial maps in the hippocampus, olfactory maps in the olfactory bulb, sound frequency coding by specific pathways in the auditory cortex, and imaging technologies that enabled visualization of activity in specific brain regions correlated with specific cognitive processes—all pointed toward the fact that circuitry determines function. But the direct demonstration that information in the brain is manifested in defined neuronal pathways was lacking.

That changed with the introduction of optogenetic methods in the early years of the twenty-first century. This methodology enabled the neurons activated by a specific experience to be tagged, while a different ensemble of neurons is tagged by a different experience. Reactivation of the different neuronal assemblies is associated with the specific experience that gave rise to that particular neuronal ensemble. Thus today, cell assemblies can literally be visualized in action.

While information in the brain is manifested in defined neuronal circuits, the definition of the circuitry is statistical rather than absolute. In the 1960s and '70s, E. Roy John was a vocal opponent of the notion that information resides in a deterministic, invariant circuitry, arguing instead that information is expressed in large-scale and dispersed patterns of neuronal activity definable only in statistical terms. He pointed to the fact that many, if not most, neurons display spontaneous activity and that neurophysiological activity at all levels consists of probabilistic elements. Those observations have been reiterated routinely up to the present day. Information in the brain thus does not rely upon the activation of precisely the same, invariant cellular elements with each iteration of the same information, but rather as a coherent pattern (or deviation from a baseline) of activity defined by the statistical properties of the neuronal population.

That said, John did not argue that functional localization does not exist—just that the functional pathway for any information in the brain can only be defined in statistical terms. Thus, cell assemblies are a reality, provided they are defined statistically rather than in terms of hardwired, invariant circuits. That the granular detail of thoughts and images can arise with such precision from a statistically variable neural substrate lingers as a fact not fully explained.

Neuronal circuitries (hence patterns of neuronal activity) are plastic. The organization of the brain, including its circuitry (what hooks up with what) is determined by genetically programmed developmental trajectories. Once in place, those circuits provide an infrastructure for the information processing that determines how an animal perceives its environment and behaves within it. But the very fact that learning can occur indicates that patterns of neuronal activity can change. And the pattern changes are now known to arise from both anatomical and functional alterations.

Anatomical plasticity was first claimed in a crude fashion by the pioneering work of Rosenzweig, Bennett, and their colleagues at the start of the 1960s. Until that time, the brain was presumed to become static and anatomically invariant once development was complete. No new neurons were believed to arise, and overall connectivity was presumed to be stable. That assumption of stability in morphology and circuitry was one of the concepts that made storage of experiential information in a chemical code, like genetic information, seem possible.

We now know that neuronal circuitry is not hardwired and invariant for the life of the organism. Circuitry is plastic because cell populations, cellular processes, and synaptic activity are all plastic.

First, cell populations can change through the loss of old cells or generation of new ones. In the early 1990s, programed cell death, which was known to occur in other organs, was demonstrated for neurons in the brain. By the early 2000s, convincing evidence had accrued for the birth of new neurons in the brains of adult animals—demonstrated first in the hippocampus, but later in other regions as well. While turnover of satellite cells in the brain, like neuroglia, astroglia, and astrocytes, was known, the demonstration that existing neurons could die and that new ones could be born in the mature brain showed that brain cell circuitry is inherently dynamic.

Second, neuronal circuitry is plastic because the structure of neuronal processes is alterable. By the start of the twenty-first century, many researchers had demonstrated that behavioral manipulations, such as exposure to enriched environments and other, more precisely controlled learning situations, could bring about changes in dendritic branching patterns and the formation of new synaptic connections. That new synapses can form while others disintegrate under the influence of different behavioral treatments—not to mention exposure to different drugs and hormones—is now a common observation.

Thirdly, synaptic activity is functionally plastic at the cellular level. Following his demonstration in 1962 that ganglia from *Aplysia* when activated appropriately show synaptic properties analogous to behavioral habituation, sensitization, and conditioning, Kandel and his colleagues in the ensuing years demonstrated the same phenomena in the intact organism. Eventually, they were able to map out a sequence of specific biochemical mechanisms for alterations in synaptic activity. Analogous research on mammals, starting with tissue slices but proceeding to measurements from intact animals, led to the discovery in the mid- to late-1970s that postsynaptic responses could be altered by the pattern of incoming excitation from the presynaptic cells, leading either to long-lasting potentiation or depression. The mechanisms for these phenomena, which immediately suggested a mechanism for learning at the cellular level, were worked out during the final decades of the twentieth century. Today, the reality of functional plasticity at the cellular level of the nervous system, even absent obvious morphological changes, is an established fact.

The pattern of projection from a neural region to its target region is labile and depends on functional stabilization through use. By the late 1970s, ingenious surgical manipulations (mainly in fishes, amphibians, and the chick) had shown that retinotectal projections could rearrange themselves in adult organisms, forming new functional connections.

Comparable experiments were eventually conducted in mammals as well, showing the ability of neural circuits to reorganize following brain damage. In rodents and primates, similar experiments involving lesions in specific brain regions that caused rerouting or reorganization of neuronal projection pathways from those regions to other brain areas have substantiated the view that neuronal pathways in the brain are genetically determined during development, but become stabilized in early life by recurrent activation, and may be reorganized later in life in subtle ways due to patterns of use or in major ways due to injury.

Kandel and Spencer in 1968 wrote about the relevance of developmental studies to mechanisms of learning.[1] This was the idea that drew me toward research on neural development, as a compliment to my interest in mechanisms of memory. To the extent that neuronal pathways are both stabilized during development but labile thereafter, the mechanisms involved could shed light on how memory circuits are established and preserved over time. The dramatic rise in synaptogenesis early in development, followed by a loss of synaptic connections and cells over time, is indeed mirrored by similar phenomena to a less dramatic degree during learning. Mechanisms of memory deposition may thus be viewed as lying on a continuum with mechanisms of neuronal circuit formation arising during development. This is not to say, however, that all the mechanisms that determine initial connectivity are necessarily involved in the reorganization of connectivity during memory formation (see below).

Mechanisms of memory involve both a thing in a place *and a* process in a population. Even when casting the issue in such starkly contrasting terms, Roy John did not assert that they were mutually exclusive.[2] While his emphasis on information as a transactional process in large populations of cellular activity led him to emphasize "a process in a population," he realized that any lasting change in those population dynamics must involve changes at (presumably a large number of) finite loci in the brain.[3] No single change at a specific site could represent the encoding of memory, since the informational content of memory necessarily required the dynamic activity of a large population of brain cells, but specific changes at finite sites constituted, at a minimum, a collection of "certain things in many places."

Neural plasticity involves multiple molecular changes. The title of my doctoral dissertation in 1969 was "Biochemical Changes in Stimulated Brain and Mechanisms of Information Processing." The meaning of *changes, stimulation,* and *information processing* was intentionally broad in all three cases. By that point, it was already clear that information processing, which could mean any aspect of dynamic brain activity, was likely to involve multiple transmitter systems, possibly modulated by a host of newly discovered molecular species, from peptides to hormones, to secretory products from nonneuronal brain cells to nutrients and other blood constituents.

As the reality of ultrastructural changes in brain circuitry became apparent—from neurogenesis to synaptogenesis, to growth and retraction of dendritic branches and axons—the need for an appropriate sequence of biochemical events underlying the structural modifications became apparent. The idea proposed in my dissertation and promoted by Robin and me at Wayne State, to search for a coherent profile of biochemical correlates of brain plasticity, was conceptually sound and in keeping with everyday observations in the neurochemical research labs at the time. Our problem was that the concept was ahead of the capability of the available contemporary techniques.

The advent of new gene expression technology, including the use of polymerase chain reactions to amplify tiny amounts of DNA in the mid-1980s, and microarray platforms to screen for thousands of gene expression products at a time in the 1990s, significantly narrowed the gap between concept and technical capability. In general, those technological advances revealed a complex pattern of gene expression associated with behavioral plasticity. It became apparent that even for neuronal circuits that are not altered anatomically, changes in the function of different elements of the circuitry occur under a variety of conditions. Using DNA microarray analysis, I was able to estimate that 2.7 percent of genes are differentially expressed in the brains of rats subjected to a brief but intense behavioral stimulation.[4] Mine was one of many labs during the first decade of this century that began to provide a quantitative estimate of the number of different biochemical changes—the number of needles in the haystack—associated with information processing in the brain.

David Sweatt has drawn attention to a conceptual problem seldom discussed: namely, the discordant time course between metabolic turnover of all molecules (which varies from seconds to weeks) and the persistence of long-term memories (which can last for a lifetime).[5] If memory is indeed ultimately reliant on the molecular building blocks of brain cells, there has to be a way for molecules involved in neural plasticity to perpetuate themselves. Too little attention has been payed to this problem.

What We Are Still Uncertain About

While establishment of connections between one neural region (such as the retina) and a target region (such as the tectum) depends on chemical information, the nature of that information is only partially understood. The concept that chemical patterns distributed in gradient fashion within target regions of the brain govern the orderly pattern of axonal input and synaptogenesis to that region has existed since the midtwentieth century. While an abundance of molecules as candidate codes for positional information have been discovered,[6] proof that they do in fact provide positional information during development has been elusive.

Starting with the work of my first graduate student, Bruce Gray, and continuing with research I carried out with Robin at Wayne State and Carol later at the Shriver Center, one of my major objectives was to discover one or more of those chemical gradients. My work was focused on gradients of glycoproteins and gangliosides, much as Barondes had envisioned.[7] Carol and I demonstrated that gradients do exist, both within the hippocampus of rats and the optic tectum of chick embryos, though they oscillate; and complex neurogenetic patterns in dynamic flux during the same periods are also occurring, which obscures the meaning of any chemical gradients.[8] During this same period, work by Urs Rutishauser and his colleagues, beginning in Gerald Edelman's lab, was demonstrating the presence of a neural cell adhesion molecule (NCAM),[9] the first of what would turn out to be a family of adhesion factors important for proliferation of neuronal processes and the formation of synaptic contacts. The assumption that NCAMs and other adhesion factors are mobilized in a gradient fashion is not unreasonable and remains an active area of investigation.

The topological projection of one neural population to another involves multiple complex processes, which probably cannot be explained merely by chemospecific tags, but chemoaffinity explanations for the ability of ingrowing neural processes to recognize the appropriate target cell to which they should attach remain logically attractive, although largely unproven.[10]

The extracellular space among brain cells probably influences the activity of neuronal patterns of activity, but does so in as yet unknown ways. The brain cell microenvironment (BCM), as I initiated the term in the report of the work session on the subject at NRP,[11] is the extracellular space surrounding all neurons and glia. It consists of an ionic solution of mainly Na^+, Ca^{+2}, and Cl^- with the approximate composition of the cerebrospinal fluid. It therefore provides a channel for electrical conduction into which project protein receptors and polyanionic (negatively charged) molecules, like glycosaminoglycans, glycoproteins (like NCAM), and gangliosides.[12] It forms a liquid bridge between neurons and glia, and a conduit for current flow among localized regions of brain cells, as envisioned in Adey's tricompartmental model of brain function.

Development of diffusion tensor imaging (DTI) in the 1990s provided a way to visualize changes in the diffusible properties of water in highly localized regions of the intact brain.[13] Changes in the diffusibility of water could reflect changes in the extracellular matrix due to structural tissue remodeling[14] or alterations in the nature or density of specific molecular patterns on neuronal or glial cell surfaces. DTI measurements consistent with this type of plasticity have been reported to accompany short-term learning in rats[15] and language learning in humans.[16] Whether altered conductivity in the extracellular space around neurons and glia actually carries information salient to brain function, or influences the flow of information through neuronal circuitry, or is an incidental side-effect of neuronal (and possibly glial) electrophysiological activity with no functional importance, is still not known.

Nonneuronal cells probably influence patterns of brain activity, but do so in as yet unknown ways. Neuroglial cells are nonneuronal cells that make up the majority of cells in the brain. They fall into three general categories. *Microglia* are small cells that migrate to sites of injury or inflammation. *Oligodendroglia* form the myelin sheath that surrounds and insulates the axons of central neurons. *Astrocytes* appear throughout the brain, surrounding and enmeshed among neuronal processes with which they are intimately associated. They are thought to be involved in the buffering of extracellular pH and ion concentrations, particularly of K^+, the synthesis of neurotransmitters and their precursors, and the transfer of nutrients from the blood to nerve cells.[17]

As long ago as the 1950s, Holger Hydén suggested that neurons and neuroglia act as a functional unit in the brain.[18] Given the functions for glia itemized above, it would be surprising if glia and neurons did *not* interact functionally. Numerous clinical studies have cited abnormal densities of astrocytes or their identifying molecule, glial fibrillary acidic protein, associated with psychological disorders;[19] but these could simply reflect deficits in the supportive roles that glia play for neurons. Whether any neuroglia play other than a passive role in brain activity is a question still under active study. Examples of a striking capacity for adult astrocytes to undergo reversible morphological changes in response to stimuli, which enhance neuronal activity have been reported.[20] Ectopic expression of an NCAM by astrocytes in mice appears to be linked to increased behavioral flexibility and selectivity while learning and relearning.[21] DTI measurements as described above have been interpreted to suggest that microstructural changes in the hippocampus and parahippocampus can occur after only two hours of training.[22]

Neuroglia are known to secrete cytokines, which have been shown to affect behavior. Interleukin (IL)-1 and IL-6, for example, inhibit activity, suppress appetite, inhibit sexual behavior, and modulate the action of catecholamine neurotransmitters.[23] These changes are particularly pronounced in disease states, which makes sense because cytokines traditionally are known to mediate immunological responses to infection. However, research from my lab has suggested that some cytokines could play a role in normal brain function as well.[24]

What Can Probably Be Ruled Out

There is no evidence that the qualitative content of memory is encoded chemically, apart from the circuitry that mediates the content. During the 1960s when discovery of the structure of DNA was only a decade old and new mechanisms of genetic information storage and readout were being uncovered on a weekly basis, the encoding of experiential information in molecular structures was not inherently implausible. Revelation of the biochemical nature of immunological memory did nothing to dispel the possibility either. Today, however, the storage of the qualitative content of memory in a strictly chemical form seems a very remote possibility, for which there is no evidence.

The concept that chemical changes in the brain must play a role in memory storage in some way has always been and continues to be assumed. What no longer is considered plausible is that a correspondence exists between any molecular structure and any qualitative content of memory: no molecular structure uniquely encodes the color red or your grandmother's face.[25] Following the early versions of theoretical speculation on chemical coding of memory from the turn of the twentieth century up to about 1960, even the pioneers of research on chemical correlates of memory did not attribute an informational correspondence between molecular structures and qualitative memory content. This included Georges Ungar, the most persistent advocate for interanimal chemical transfer of memory. A careful reading of Ungar's views reveals a belief that memory transfer was really just transfer of tags for complex circuits mediating the neural plasticity that occurs during learning.[26]

Indeed, the demise of any notion that molecular structure encodes qualitative memory has been due mainly to the emergence of more plausible alternatives—especially broad agreement that cell assemblies are a reality (see above) and that information resides in spatiotemporal patterns of large ensembles of neurons. Any structural changes in circuitry—growth of new dendritic spines, formation of new synapses, or resculpturing of the cell surface of either neurons or neuroglia, for instance—will necessarily involve a change in the turnover of certain classes of molecules. Such molecules, however, would contribute to the content of the engram in the same way that a brick contributes to the structure of a building, with no meaning in isolation other than the extent to which it is part of the form of the entire edifice

1   Kandel, E. R. and W. A. Spencer. 1968. Cellular neurophysiological approaches in the study of learning. *Physiol Rev* 48: 65–134.

2   John, E. R. 1967. *Mechanisms of memory*. New York Academic Press.

3   John, E. R. 1972. Switchboard versus statistical theories of learning and memory. *Science* 177: 850–851.

4   Irwin, L. N. 2001. Gene expression in the hippocampus of behaviorally stimulated rats: analysis by DNA microarray. *Brain Res Mol Brain Res* 96: 163–9.

5   Sweatt, J. D. 2003. *Mechanisms of Memory*. 1st ed. San Diego: Academic Press.

6   Sharma, S. C. 1999. Retinotectal interactions. In *Encyclopedia of Neuroscience*, edited by G. Adelman and B. H. Smith. Amsterdam: Elsevier; pp. 1804–6.

7   Barondes, S. H. 1970. Brain glycomacromolecules and interneuronal recognition. In *The Neurosciences: Second Study Program*, edited by F. O. Schmitt. New York: The Rockefeller University Press.

8   Irwin, L.N. and C.C. Irwin. 1979. Developmental changes in ganglioside composition of hippocampus, retina, and optic tectum. *Develop Neurosci* 2: 129–138; Irwin, L. N. and C. C. Irwin. 1982. Developmental changes and regional variation in the ganglioside composition of the rat hippocampus. *Brain Res* 256: 481–5.

9   Rutishauser, U., J. P. Thiery, R. Brackenbury, and G. M. Edelman. 1978. Adhesion among neural cells of the chick embryo. III. Relationship of the surface molecule CAM to cell adhesion and the development of histotypic patterns. *J Cell Biol* 79: 371–81. NCAM is a glycoprotein consisting of multiple sialic acid residues.

10   Sharma, S. C. 1999. Retinotectal interactions. In *Encyclopedia of Neuroscience*,

edited by G. Adelman and B. H. Smith. Amsterdam: Elsevier; pp. 1804–6.

[11] Schmitt, F.O. and F. Samson. 1969. Brain cell microenvironment. *NRP Bull* 7: 301–373.

[12] Nicholson, C. 1999. Brain cell microenvironment. In *Encyclopedia of Neuroscience*, edited by G. Adelman and B. H. Smith. Amsterdam: Elsevier.

[13] Basser, P. J. 1995. Inferring microstructural features and the physiological state of tissues from diffusion-weighted images. *NMR Biomed* 8: 333–44.

[14] Sagi, Y., I. Tavor, S. Hofstetter, S. Tzur-Moryosef, T. Blumenfeld-Katzir, and Y. Assaf. 2012. Learning in the fast lane: new insights into neuroplasticity. *Neuron* 73: 1195–1203.

[15] Hofstetter, S. and Y. Assaf. 2017. The rapid development of structural plasticity through short water maze training: A DTI study. *Neuroimage* 155: 202–208.

[16] Hofstetter, S., N. Friedmann, and Y. Assaf. 2017. Rapid language-related plasticity: microstructural changes in the cortex after a short session of new word learning. *Brain Struct Funct* 222: 1231–1241.

[17] Somjen, G. G. 1999. Glial cells: functions. In *Encyclopedia of Neuroscience*, edited by G. Adelman and B. H. Smith. Amsterdam: Elsevier.

[18] Hydén, H., ed. 1959. *Biochemical changes in glial cells and nerve cells*. Edited by F. Brücke. Vol. III, pp. 66–89, *Fourth Intl. Congress of Biochemistry*; Hydén, H. and A. Pigon. 1960. A cytophysiological study of the functional relationship between oliodendroglial cells and nerve cells of Deiters' nucleus. *J Neurochem* 6: 57–72.

[19] Webster, M. J., M. B. Knable, N. Johnston-Wilson, K. Nagata, M. Inagaki, and R. H. Yolken. 2001. Immunohistochemical localization of phosphorylated glial fibrillary acidic protein in the prefrontal cortex and hippocampus from patients with schizophrenia, bipolar disorder, and depression. *Brain Behav Immun* 15: 388–400; Webster, M. J., M. B. Knable, N. Johnston-Wilson, K. Nagata, M. Inagaki, and R. H. Yolken. 2001. Immunohistochemical localization of phosphorylated glial fibrillary acidic protein in the prefrontal cortex and hippocampus from patients with schizophrenia, bipolar disorder, and depression. *Brain Behav Immun* 15: 388–400; Webster, M. J., J. O'Grady, J. E. Kleinman, and C. S. Weickert. 2005. Glial fibrillary acidic protein mRNA levels in the cingulate cortex of individuals with depression, bipolar disorder and schizophrenia. *Neuroscience* 133: 453–61; Si, X., J. J. Miguel-Hidalgo, G. O'Dwyer, C. A. Stockmeier, and G. Rajkowska. 2004. Age-dependent reductions in the level of glial fibrillary acidic protein in the prefrontal cortex in major depression. *Neuropsychopharmacology* 29: 2088–96.

[20] Theodosis, D. T. and D. A. Poulain. 1999. Contribution of astrocytes to activity-dependent structural plasticity in the adult brain. *Adv Exp Med Biol* 468: 175–82.

[21] Wolfer, D. P., H. M. Mohajeri, H. P. Lipp, and M. Schachner. 1998. Increased

flexibility and selectivity in spatial learning of transgenic mice ectopically expressing the neural cell adhesion molecule L1 in astrocytes. *Eur J Neurosci* 10: 708–17.

[22] Sagi, Y., I. Tavor, S. Hofstetter, S. Tzur-Moryosef, T. Blumenfeld-Katzir, and Y. Assaf. 2012. Learning in the fast lane: new insights into neuroplasticity. *Neuron* 73: 1195–1203.

[23] Dunn, A. J. and J-P. Wang. 1999. Cytokines and the brain. In *Encyclopedia of Neuroscience*, edited by G. Adelman and B. H. Smith. Amsterdam: Elsevier.

[24] Irwin, L. N. and D. M. Byers. 2012. Novel odors affect gene expression for cytokines and proteinases in the rat amygdala and hippocampus. *Brain Res* 1489: 1–7.

[25] But the possibility that single *cells* could encode complex memories has not been totally ruled out. The existence of such cells, generically referred to as "grandmother neurons," is widely rejected, though prematurely so in the view of J. S. Bowers, N. D. Martin, and E. M. Gale. 2019. Researchers keep rejecting grandmother cells after running the wrong experiments: The issue is how familiar stimuli are identified. *Bioessays* 41: e1800248.

[26] Ungar, G. 1968. Molecular mechanisms in learning. *Perspect Biol Med* 11: 217–32; Ungar, G. 1973. The problem of molecular coding of neural information. A critical review. *Naturwissenschaften* 60: 307–12.

Afterword

Ross Adey pursued a vigorous program of research for over two decades after his promotion to emeritus status at the Neurosciences Research Program. Controversial until his death in 2004, his research in later years was devoted to drawing attention to the unknown, long-term consequences of exposure to environmental electromagnetic fields. In 1988, he returned to a central theme of his career, which had paralleled my own views that "calcium ions play a key role in [neural] stimulus amplification, probably through highly cooperative alterations in binding to surface glycoproteins."

Roy John was never a sought-after scientist at NRP, but should have been. His publication of *Mechanisms of Memory*, in 1967, set a standard for depth and insight into the problem unmatched till David Sweatt's book of the same title was published in 2003. Well into the twenty-first century, John was still pursuing his devotion to neurometrics (quantitative electroencephalography), medical instrumentation development, clinical applications of his research, and theories of consciousness. He died in 2009, having never engaged in the debate that Ted Bullock had sought with him to no avail through the offices of NRP.

Samuel Barondes spent many years at the University of California, San Diego, studying mainly cell aggregation in the slime mold as a model for intercellular recognition. In 1986, he became chair of psychiatry at the University of California, San Francisco School of

Medicine, devoting much of his time to writing books for the general public about psychiatric genetics, psychopharmacology, and personality.

Mark Rosenzweig and Edward Bennett remained at Berkeley, making important contributions to the neuropharmacology of learning and memory until their retirements in the midnineties. They and their research partner, Marian Diamond, continued to write about brain plasticity into the first decade of the twenty-first century. Rosenzweig died in 2009, Diamond in 2017, and Bennett in 2018.

Bernard Agranoff became one of the most widely known and respected figures in the field of neurochemistry. He stayed active at the University of Michigan, extending his early groundbreaking work on lipid metabolism and making important contributions to the study of retinotectal specificity. Unknowingly, he coined the title of this book.

James V. McConnell remained at the University of Michigan for his entire career. After his research program diminished, he wrote a widely adopted textbook in psychology that made him rich. He died in 1989.

Frank Schmitt turned over leadership of the Neurosciences Research Program to Gerald Edelman in 1980, who moved the organization to the Rockefeller University campus in New York, then to San Diego. The NRP was transformed in the process and became progressively less influential as neuroscience research proliferated in too many directions with too much force to be influenced by any one figure or institution. Schmitt retired to an office at MIT where he continued to think and write about neuroscience until the morning he died in 1994.

Fred Samson moved from the Lawrence campus of the University of Kansas to the Ralph Smith Center for Mental Retardation at the medical campus in Kansas City in 1972. In 1990, he stepped down from that position, demonstrating at his retirement dinner the one-armed hand stand he had perfected as an acrobat in his youth. He retained his office and continued to work vigorously every day, in concert with Frank Schmitt until Schmitt's death, then on his own, pursuing especially questions related to the role of free radicals in biological systems. He continued to work daily on the scientific questions that were his passion for a decade after his retirement until shortly before his death in 2004.

Dianna Redburn rose through the ranks at the University of Texas, Houston, and developed an international reputation for her excellent research in retinal neurochemistry. She was tempted by but turned down an opportunity to succeed Fred Samson as director of the Ralph Smith Center for Mental Retardation at the Kansas University School of Medicine and moved instead to the University of Tennessee Health Science Center in Memphis in 1997. After she and Ray divorced, she married Leonard Johnson and published under the name D. A. Johnson.

Kay and our son, Michael Sean, moved to Oregon in 1965. She eventually received a PhD in English from the University of Oregon and taught English at the University of Alberta in Edmonton for many years. She published numerous essays and textbooks before turning to fiction as the author or coauthor of a series of mystery novels. She remarried twice. Her second husband, Joe Stewart, adopted Sean. Kay, Joe, and I remained good friends. Sean grew up with a bewildering array of relatives and became a critically acclaimed author of science fiction and fantasy novels, writing under the name Sean Stewart. He and colleagues at the company they cofounded, Fourth Wall Studios, won an Emmy for interactive television in 2013. He is recognized as one of the founders of the twenty-first-century genre of interactive and alternate reality entertainment.

Penny May Hopkins taught briefly in New Jersey after completing her postdoctoral years at the Museum of Natural History in New York, then moved to the University of Oklahoma in the late 1970s where she became the first woman to be a tenured professor in the Department of Zoology on the strength of significant discoveries in molecular endocrinology. Upon her retirement, she pursued her deep concern about the destructive effects of petroleum-based pollutants, publishing a book on the subject in 2019.

Dorothy Haecker earned her PhD at the University of Kansas and taught at colleges in Missouri and California before returning to San Antonio for twenty-four years as a professor at Palo Alto College, where she was acclaimed for her teaching excellence. She retired as a professor emeritus of philosophy in 2014.

Bob Godbout taught philosophy at the University of Evansville for a few years, and Muriel obtained a degree in sociology while they were there. Bob then decided to become a protestant minister. He studied at the Princeton School of Divinity (finally acquiring that Ivy League degree) and subsequently served congregations in New Jersey, California, and New York, before he and Muriel retired to their childhood home of Manchester, New Hampshire.

Jerry Mitchell remained in the Department of Anatomy at the Wayne State University School of Medicine and became one of Detroit's most active community leaders in the arts and historical preservation. He and Marilyn worked tirelessly to maintain the Boston-Edison neighborhood as a model integrated inner-city community. They bought Henry Ford's house on Edison Avenue and restored it to the state it was in when Ford started assembling automobiles in the garage on the property. Marilyn eventually went to law school and became a partner in a prestigious downtown Detroit law firm.

Bruce Gray, my first graduate student, took a long and winding path to a PhD, which he obtained eventually from the University of Rhode Island. He carried out research at the University of Connecticut for several years, then moved to Simmons College himself in becoming one of the most popular professors in the Department of Biology until his retirement in 2018.

Dan Michael obtained both an MD and PhD from Wayne State University with Jerry Mitchell as his major advisor. He became a well-known neurosurgeon in Detroit, expert especially on the physiological and biochemical responses to brain trauma. During another short-term visit to my lab at the University of Texas at El Paso, he collaborated with Donna Byers and me on a study of gene expression responses to brain trauma, which became one of my most highly cited publications. He was a founding member of the Inverted Skull Society, named after a surgical procedure for which he became famous among a select group of his friends.

Marc Abel completed his PhD at the University of Texas, Houston, and worked in the pharmaceutical industry for several years, learning techniques of molecular biology that he later put to use in an active research program as a faculty member at the Chicago Medical College—later renamed the Rosalind Franklin University of Medicine and Science. Eventually, he became dean of the University's College of Pharmacy. He, too, was a founding member of the Inverted Skull Society.

David Terrian received a PhD from Wayne State University, then worked for a number of years as a research scientist at the Brooks School of Aviation Medicine in San Antonio prior to becoming a faculty member at the East Carolina University School of Medicine. There he established a productive research program in neurochemistry, which turned later to equally productive cancer research.

Robin Barraco received tenure the year after I left Detroit. In place of the grant he gave up for my sake, he would later get larger ones on his own. His career flourished as he became an expert with an international reputation on the nucleus solitarius of the brain stem. He eventually remarried and had a daughter. At the height of his career and accomplishments, he developed bone cancer and died far too young at the age of forty-nine, spirited and controversial among his colleagues but generous with me to the end.

I spent twelve enjoyable years as a faculty member in the Department of Biology at Simmons College (now Simmons University). In 1983, I was elected to the Board of Aldermen in the city of Newton and was reelected for three successive terms after that. Carol worked as a biochemist in several labs in the Boston area after leaving the Shriver Center in 1982. In 1989, she received an MBA from the School of Management at Simmons College, then worked briefly for an international pharmaceutical firm until a lull in the industry in 1991 enabled me to talk her into moving with me to El Paso. Our sons, Anthony and Brian, grew up to make us proud, but in pursuits unrelated to science. I served as chair of biological sciences at the University of Texas at El Paso until 1999, just as new technologies for the analysis of gene expression burst onto the scene. In the closing month of the millennium, I performed my first experiment on animal behavior in over twenty years, using DNA microarray analysis to return to the search for a coherent sequence of molecular events related to the processing and storage of information in the brain.

Index

LeBaron, Francis, 172
Lees, Marjorie B., 169, 247, 339, 365–66
Li, C. H., 267
Lieberman, Irving, 210
lipotropin (LPH), 266
Liss, Samuel, 234
Loehr, Karen, 365, 367
Lolley, Richard, 171, 174
Lømo, Terje, 325
long-term potentiation (LTP), 281, 325, 376
Lopilato, Jane, 367
Lubbock, Texas, 36
Lund, Ray, 352
Lynch, Gary, 281, 322, 324–25

M

Malin, David, xiv
Malkin, Leonard, 271–72
Mancini, Judy, 214, 217, 220, 223, 234
Markert, Clement, 49
Massachusetts General Hospital, 29, 169
Matthaei, Heinrich, 22, 27–28, 31n18
May, Penny, 37, 83, 105–6, 188, 199–200, 268, 304, 368, 389
McCarthy, Eugene, 102, 153, 160
McCluer, Robert, 206–7, 264, 286–87, 348, 360
McConnell, James V., xiv, 53–54, 87–91, 97n1, 97n8, 119, 388
McGaugh, James L., xiv, 56–57, 58n14, 83, 107, 152–53, 182
McGovern, George, 230
McIlwain, Henry, 123, 169
McLean Hospital, 169
McMichael Canadian Art Collection, 283
melanocyte stimulating hormone (MSH), 266
Melnechuk, Ted, 155
memory

biochemistry of, 2, 158, 293
disruption of, 83, 87, 127
encoding of, 35, 50, 87, 187, 324, 377
immunological, 7, 20, 23, 30, 382
long-term (LTM), 57
mechanisms of, 2, 9, 13, 15–16, 25, 56, 100, 110, 197, 233, 242, 323, 377, 387
neurochemistry of, 2, 273, 293–94
retrieval of, xii, 11, 50, 128
short-term (STM), 57
storage of, xi–xii, 17, 128, 296, 323, 382
switchboard theories of, 11
transfer of, 87, 107, 127, 136n3, 148, 187, 201, 204, 382
meprobamate, 53
Merck Company Foundation, 210
Merker, Philip, 215, 218
messenger RNA (mRNA), 21–22, 384
Meyer, Ronald, 337, 352
Meyerhoff, Otto, 60, 119
Michael, Daniel, 287, 349, 355, 357, 390
Michael, Melinda. See Raab, Melinda
Michaelis, Leonor, 49
Microglia, 381
Miltown, 53
Mitchell, Jerry, 101–2, 114–15, 125, 131, 133, 161–62, 166, 173, 191, 223, 231, 287–88, 368, 390
Mitchell, Marilyn, 114, 125, 131–33, 147, 151, 154, 159–60, 162, 166, 173, 191, 287–88, 368, 390
Moller, Fred, 49
Monday Night Football, 197–98
Monod, Jacquez, 21
moon landing, 177, 251
Moore, Robert, 313–14
Moore, Stanford, 272
Morgan, Barry, 265–66

morphine, 84, 95–96, 108, 224–26, 264, 332

Morris, Howard, 265–66

Moscow, 7

Mountcastle, Vernon, 30, 299–301, 308, 313, 315–16, 318, 342–44, 346, 350

Murano, Gene, 243, 277

N

NASA, 16, 121–23, 177, 183, 190

neural cell adhesion molecule (NCAM), 312, 380–81, 383n9n9

neuroglia, 39, 376, 381, 383

neuron doctrine, 10

Neurosciences Research Foundation, 64

Neurosciences Research Program
    Associates of, 63, 65, 157, 180, 294, 298–301, 307, 310, 315–16, 330, 340, 343, 346
    formation of, 59, 62, 67
    ISP of 1966, *66*
    ISP of 1969, *175*
    ISP of 1977, *298*
    leadership transition of, 388
    staff scientists, 65, 301
    Stated Meetings of, 157, 180, 330–31, 340–41, 357

Neurosciences Research Program Bulletin
    Brain Cell Microenvironment, 159, 164, 175, 278, 293, 302
    origin, 64
    Retinotectal Connectivity, 302, 335, 337, 352, 354, 359

Newton, Massachusetts, 334, 360, 369, 391

Nickerson, Richard, 367

Nirenberg, Marshall, 22, 27–28, 31n18, 51, 67

Nixon, Richard, 160–61, 165, 189, 216, 220, 224, 230–32, 244, 251–52

NRP Bulletin, 64–65, 175, 302, 335, 337, 352, 354, 359

O

Ochoa, Severo, 22, 63, 67

olfactory maps, 374

opiate, endogenous, 226. *See also* endorphin

opiate receptor, 224

optogenetic methods, 374

oxytocin, xi, 202–4, 267

Ozalas, Frans, 234

P

Palay, Sanford L., 313, 315

Parr, W., 219

Parsons, Kansas, 165–66, 172–73, 184, 189–91, 206

Pasternak, Gavril, 265

Pauling, Linus, 23, 26, 240

peptide, 23, 30n8, 95–96, 127, 130–31, 135, 187, 201, 203–5, 226–27, 265, 267, 279, 281, 296–97

Pert, Candace, 225–26, 294

Perutz, Max, 26

Pevzner, Leonid, 83

phase sequence, 12

Piperberg, Joel, 367

planaria, 39, 54, 88, 90, 97

plasticity
    anatomical, 375
    of behavior, 181, 378
    brain, xii, 375, 378, 388
    cerebellar, 326
    of circuitry, 11, 376, 382
    connectionistic theories of, 117
    functional, 263, 338, 375–76
    long-term, 327
    mechanisms of, xi, 15

Ingram Content Group UK Ltd.
Milton Keynes UK
UKHW011837230523
422246UK00011B/80/J